Legal Notice

For information on bulk purchases and licensing agreements, please email

support@SATPrepGet800.com

ISBN-13: 978-1-951619-04-6

Also Available from Dr. Steve Warner

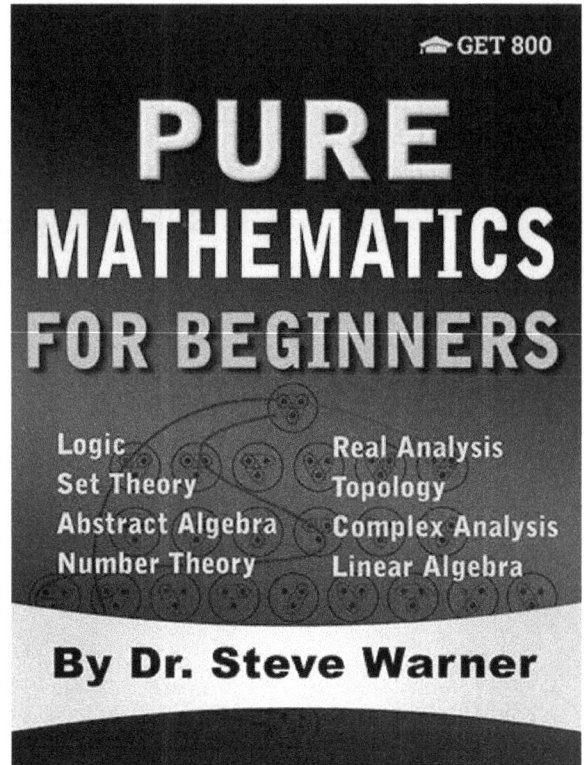

CONNECT WITH DR. STEVE WARNER

www.facebook.com/SATPrepGet800

www.youtube.com/TheSATMathPrep

www.twitter.com/SATPrepGet800

www.linkedin.com/in/DrSteveWarner

www.pinterest.com/SATPrepGet800

Also Available from Dr. Steve Warner

CONNECT WITH DR. STEVE WARNER

www.facebook.com/SATPrepGet800

www.youtube.com/TheSATMathPrep

www.twitter.com/SATPrepGet800

www.linkedin.com/in/DrSteveWarner

www.pinterest.com/SATPrepGet800

Set Theory
for Pre-Beginners

An Elementary Introduction to Sets, Relations,
Partitions, Functions, Equinumerosity, Logic,
Axiomatic Set Theory, Ordinals, and Cardinals

Dr. Steve Warner

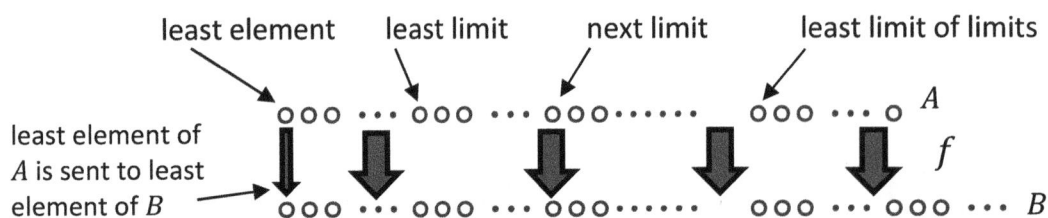

least element   least limit   next limit   least limit of limits

least element of
A is sent to least
element of B

A

f

B

Table of Contents

SET THEORY FOR PRE-BEGINNERS

Shortly after the release of *Set Theory for Beginners,* I began receiving messages from a wide range of readers that were enjoying the book. However, I quickly learned that the book was too difficult for some readers. In particular, many high school teachers mentioned that the more theoretical aspects of the book were too much of a challenge for many of their students. I refer to these students/readers as "pre-beginners." By a "pre-beginner," I mean a student that is ready to start learning some more advanced mathematics, but is not quite ready to dive into proofwriting.

The book you are currently reading was written with these "pre-beginners" in mind. It provides a more basic **nonrigorous** introduction to set theory. The rigor presented in *Set Theory for Beginners* has been replaced by a more elementary approach to the topics.

There are no prerequisites for this book. The content is completely self-contained. Furthermore, reading this book will naturally increase a student's level of "mathematical maturity." Although there is no single agreed upon definition of mathematical maturity, one reasonable way to define it is as "one's ability to analyze, understand, and communicate mathematics." A student with a very high level of mathematical maturity may find this book very easy—this student may want to go through the book quickly and then move on to *Set Theory for Beginners*. A student with a lower level of mathematical maturity will probably find this book more challenging. However, the reward will certainly be more than worth the effort.

If you read this book and complete the exercises along the way, then your level of mathematical maturity will continually be increasing. This increased level of mathematical maturity will not only help you to succeed in advanced math courses, but it will improve your general problem solving and reasoning skills. This will make it easier to improve your performance in college, in your professional life, and on standardized tests such as the SAT, ACT, GRE, and GMAT.

At the end of each lesson there is a Problem Set. The problems in each of these Problem Sets have been organized into five levels of difficulty, followed by several Challenge Problems. Level 1 problems are the easiest and Level 5 problems are the most difficult, except for the Challenge Problems. If you want to get just a small taste of pure mathematics, then you can work on the easier problems. If you want to achieve a deeper understanding of the material, take some time to struggle with the harder problems.

The author welcomes all feedback. Feel free to email Dr. Steve Warner with any questions and comments at

steve@SATPrepGet800.com

LESSON 1
SETS AND SUBSETS

Describing Sets Explicitly

A **set** is simply a collection of "objects." These objects can be

- numbers such as 3, 7, or 16.

- letters such as a, b, h, or z.

- colors such as blue, red, or green.

- shapes such as △, □, or ○.

- birds such as duck, flamingo, sparrow, or ostrich.

- ... or just about anything else you can imagine.

We will usually refer to the objects in a set as the **members** or **elements** of the set.

If a set consists of a small number of elements, we can describe the set simply by listing the elements in the set inside curly braces, separating elements by commas. We call this method of describing a set the **roster method**.

Example 1.1:

1. {butterfly, grasshopper} is the set consisting of two elements: *butterfly* and *grasshopper*.

2. {minotaur, beetle, fireman, boat, screwdriver, lithium} is the set consisting of six elements: *minotaur*, *beetle*, *fireman*, *boat*, *screwdriver*, and *lithium*.

3. $\{0, 2, 5, 9, 23\}$ is the set consisting of five elements: 0, 2, 5, 9, and 23. The elements in this set happen to be *numbers*.

Exercise 1.2: Determine how many elements are in each of the following sets and then list the elements in the set.

1. $\{j, k, t, u, v\}$ _____

2. {triceratops, tulip, lima bean} _____

3. $\{1.6, 5.66, 9.03, 15.27\}$ _____

4. {Earth, Mars, Neptune, Venus, Jupiter, Mercury} _____

5. {oxygen, helium, nitrogen, sulfer} _____

A set is determined by its elements and not the order in which the elements are presented. For example, the set $\{2, 0, 3, 1, 4\}$ is the same as the set $\{0, 1, 2, 3, 4\}$.

Also, the set $\{0, 0, 1, 2, 2, 2, 3, 3, 4\}$ is the same as the set $\{0, 1, 2, 3, 4\}$. If we are describing a set by listing its elements, the most natural way to do so is to list each element just once.

We will usually name sets using capital letters such as A, B, C,..., and so on. For example, we might write $A = \{a, b, c, d\}$. So, A is the set consisting of the elements a, b, c, and d.

Example 1.3: Consider the sets $X = \{0, 1\}$, $Y = \{1, 0\}$, $Z = \{0, 1, 0, 1, 0\}$. Then X, Y, and Z all represent the same set. We can write $X = Y = Z$.

Exercise 1.4: For each of the following, circle the set that is **not** equal to the others.

1. $\{5, 7, 9\}$, $\{7, 5, 9\}$, $\{3, 3, 7, 9\}$, $\{5, 5, 7, 9, 9\}$

2. $\{a, x, t, v, d, b\}$, $\{a, a, x, x, x, t, t, t, t, v, v, v, d, d\}$, $\{t, a, x, d, v, d, x, a, t\}$, $\{t, a, x, x, v, d\}$

3. $\{\text{car}, \text{bus}\}$, $\{c, a, r, b, u, s\}$, $\{\text{bus}, \text{car}, \text{bus}\}$, $\{\text{car}, \text{car}, \text{bus}, \text{car}\}$

We use the symbol $\in$ to indicate membership. Specifically, $x \in A$ means "x is an element of A," whereas $x \notin A$ means "x is **not** an element of A." We will often simply say "x is in A," and "x is not in A," respectively. Membership is an example of a **relation**. It describes a relationship between objects.

Example 1.5: Let $D = \{y, j, 0, \beta, \uparrow\}$. Then $y \in D$, $j \in D$, $0 \in D$, $\beta \in D$, and $\uparrow \in D$. We can combine all this information into a single statement as follows: $y, j, 0, \beta, \uparrow \in D$.

Note: β (pronounced "beta") is a greek letter. It is the second letter of the greek alphabet.

Exercise 1.6: Let $Y = \{1, 8, 101, m, v, \text{jaguar}\}$. Determine if each of the following is true or false.

1. $8 \in Y$

2. $m \notin Y$

3. $\text{leopard} \in Y$

4. 6 is a member of Y

5. y is not an element of Y

6. $101, v \in Y$

7. $1, 8, 101, \text{jaguar} \in Y$

8. $a, m, v \in Y$

Describing Sets with Ellipses

If a set consists of many elements, we can use **ellipses** (...) to help describe the set. For example, the set consisting of the natural numbers between 1 and 80, inclusive, can be written $\{1, 2, 3, ..., 79, 80\}$ ("inclusive" means that we include 1 and 80). The ellipses between 3 and 79 are there to indicate that there are elements in the set that we are not explicitly mentioning.

Example 1.7: Let $K = \{0, 2, 4, ..., 30\}$. Then we have $6 \in K$ and $28 \in K$, whereas $29 \notin K$. Using the roster method, we have $K = \{0, 2, 4, 6, 8, 10, 12, 14, 16, 18, 20, 22, 24, 26, 28, 30\}$.

Exercise 1.8: Let $L = \{5, 7, 9, ..., 33\}$. Use the roster method to describe the set L.

We can also use ellipses to help describe **infinite sets**. The set of **natural numbers** can be written $\mathbb{N} = \{0, 1, 2, 3, \dots\}$, and the set of **integers** can be written $\mathbb{Z} = \{\dots, -4, -3, -2, -1, 0, 1, 2, 3, 4, \dots\}$.

Notes: (1) Some mathematicians exclude 0 from the set of natural numbers. In this book, 0 will always be included. Symbolically, $0 \in \mathbb{N}$.

(2) Notice how we use special character symbols to represent the natural numbers and integers. The characters $\mathbb{N}$ and $\mathbb{Z}$ are said to be **doublestruck** or in **blackboard bold**. In general, important sets are written using doublestruck character symbols. The natural numbers and integers are two such examples. We will see several more shortly.

Example 1.9:

1. The **even natural numbers** can be written $\mathbb{E} = \{0, 2, 4, 6, \dots\}$. We may also use the notation $2\mathbb{N}$ to describe this set. So, $2\mathbb{N} = \mathbb{E} = \{0, 2, 4, 6, \dots\}$.

2. The **odd natural numbers** can be written $\mathbb{O} = \{1, 3, 5, \dots\}$. We may also use the notation $2\mathbb{N} + 1$ to describe this set. So, $2\mathbb{N} + 1 = \mathbb{O} = \{1, 3, 5, \dots\}$.

3. The **positive integers** can be written $\mathbb{Z}^+ = \{1, 2, 3, 4, \dots\}$ or $\mathbb{N}^+ = \{1, 2, 3, 4, \dots\}$ (the positive integers and the positive natural numbers describe the same set).

Exercise 1.10: Use brackets, commas, and ellipses to describe each of the following sets:

1. The **even integers** $2\mathbb{Z}$ _____

2. The **odd integers** $2\mathbb{Z} + 1$ _____

3. The **negative integers** $\mathbb{Z}^-$ _____

Describing Sets with Properties

Another way to describe a set is to specify a certain property P that all its elements have in common. There are endless possibilities for what a property could be. For example, suppose that P is the property of being an insect. Then "mosquito" satisfies the property P, whereas "computer" does **not** satisfy the property P.

If we wish to describe a set with a property P, then we can use the **set-builder notation** $\{x | P(x)\}$. The expression $\{x | P(x)\}$ can be read "the set of all x such that the property $P(x)$ is true." Note that the symbol "$|$" is read as "such that."

The letter x in the set-builder notation is called a **variable**. The choice of x is completely arbitrary. We could have used any other letter or symbol. In other words, $\{x | P(x)\}$, $\{t | P(t)\}$, $\{\square | P(\square)\}$, and $\{? | P(?)\}$ all have exactly the same meaning. The idea here is that we substitute objects in for the variable. If that object has the given property, then it is in the set. If it does not have the given property, then it is not in the set. Sticking with our example above, let $A = \{x \mid x \text{ is a insect}\}$. Then we have mosquito $\in A$ ("mosquito" is an element of the set A) because when we substitute "mosquito" in for the variable x, we get a true statement. Indeed, a mosquito is an insect. On the other hand, computer $\notin A$ ("computer" is **not** an element of the set A) because when we substitute "computer" in for the variable x, we get a false statement. Indeed, the statement "a computer is an insect" is false.

Example 1.11: Let $B = \{x \mid x \text{ is a dessert}\}$. In words, we can describe the set B as "the set of all x such that x is a dessert." Ice cream is an element of this set because ice cream is a dessert. In other words, if we replace x by "ice cream," then we get the true statement "ice cream is a dessert." Symbolically, we can write ice cream $\in B$. Alligator is not an element of this set because an alligator is not a dessert. Symbolically, we can write alligator $\notin B$. However, we do have alligator $\in \{x \mid x \text{ is an animal}\}$. Similarly, we also have alligator $\in \{x \mid x \text{ is a reptile}\}$.

Exercise 1.12: Let $X = \{x \mid x \text{ is a word containing at least three distinct vowels}\}$. Determine if each of the following is in the set X.

1. hearth ___

2. grasshopper ___

3. fire away ___

4. incremental ___

5. apartment ___

Example 1.13: Let's look at a few different ways that we can describe the set $\{0, 1, 2, 3, 4\}$. We have already seen above that reordering and/or repeating elements does not change the set. For example, $\{1, 1, 0, 4, 3, 3, 2, 2, 2\}$ describes the same set. Here are a few descriptions using set-builder notation:

- $\{n \mid n \text{ is a natural number between 0 and 4, inclusive}\}$
- $\{n \mid n \text{ is an integer between 0 and 4, inclusive}\}$
- $\{k \mid k \in \mathbb{N} \text{ and } 0 \leq k \leq 4\}$
- $\{t \mid t \in \mathbb{Z} \wedge 0 \leq t < 5\}$
- $\{m \mid m = 0, 1, 2, 3, \text{ or } 4\}$

Notes: (1) The first expression in the bulleted list above can be read "the set of n such that n is a natural number between 0 and 4, inclusive." Recall that the word "inclusive" means that we include 0 and 4.

(2) The second expression can be read "the set of n such that n is an integer between 0 and 4, inclusive."

(3) The third expression can be read "the set of k such that k is a natural number and k is between 0 and 4, including both 0 and 4. Note that the abbreviation "$k \in \mathbb{N}$" can be read "k is in the set of natural numbers," or more briefly, "k is a natural number." We used the letter "k" for the variable here (as opposed to the letter "n" that was used in the first two expressions). Once again, we can use any variable name we like, as long as it doesn't lead to confusion.

(4) The fourth expression can be read "the set of t such that t is an integer and t is between 0 and 5, including 0 and excluding 5." Note that the abbreviation "$t \in \mathbb{Z}$" can be read "t is in the set of integers," or more briefly, "t is an integer." Also note that the symbol "$\wedge$" is called a **conjunction symbol** and it can be read as "and."

(5) The fifth expression can be read "the set of m such that m is 0, 1, 2, 3, or 4."

Exercise 1.14: Use set-builder notation to describe each of the following sets.

1. $\{0, 2, 4, 6, 8, 10, 12, 14, 16\}$ _____

2. $\{-5, -3, -1, 1, 3, 5, \ldots, 85, 87\}$ _____

3. $\mathbb{N}$ _____ .

4. $\mathbb{Z}$ _____

5. $2\mathbb{Z}$ _____

For easier readability, we may include a **bounding set** when using set-builder notation. If we wish to describe a set with a property P and a bounding set A, then the corresponding set-builder notation is $\{x \in A \mid P(x)\}$. As an example, consider the set $\{7, 8, 9\}$. Since every element of this set is a natural number, we can use the set of natural numbers, $\mathbb{N}$, as a bounding set. So, instead of writing $\{k \mid k \in \mathbb{N} \wedge 7 \leq k \leq 9\}$, we can use the friendlier notation $\{k \in \mathbb{N} \mid 7 \leq k \leq 9\}$. Notice how the bounding set appears to the **left** of the vertical line.

Example 1.15: In example 1.13, we described the set $\{0, 1, 2, 3, 4\}$ in various ways using set-builder notation. Let's look at a few more ways to do this, this time using bounding sets in the description.

- $\{n \in \mathbb{N} \mid n \text{ is between 0 and 4, inclusive}\}$

- $\{n \in \mathbb{Z} \mid n \text{ is between 0 and 4, inclusive}\}$

- $\{k \in \mathbb{N} \mid 0 \leq k \leq 4\}$

- $\{t \in \mathbb{Z} \mid 0 \leq t < 5\}$

- $\{m \in \mathbb{N} \mid m = 0, 1, 2, 3, \text{ or } 4\}$

In addition to the sets $\mathbb{N}$ (the natural numbers) and $\mathbb{Z}$ (the integers), let's look at a few more sets that will show up throughout this book.

The set of **rational numbers** is $\mathbb{Q} = \left\{\frac{a}{b} \mid a, b \in \mathbb{Z} \text{ and } b \neq 0\right\}$. In words, $\mathbb{Q}$ is "the set of quotients a over b such that a and b are integers and b is not zero." Some examples of rational numbers are $\frac{0}{2}, \frac{3}{7}, \frac{4}{11}$, and $\frac{-6}{13}$. We identify rational numbers $\frac{a}{b}$ and $\frac{c}{d}$ whenever $ad = bc$. For example, $\frac{1}{4} = \frac{2}{8}$ because $1 \cdot 8 = 4 \cdot 2$. We also abbreviate the rational number $\frac{a}{1}$ as a. In this way, we can think of every integer as a rational number. For example, we have $\frac{0}{2} = \frac{0}{1}$ (because $0 \cdot 1 = 2 \cdot 0$), and therefore, we can abbreviate $\frac{0}{2}$ as 0. Similarly, we can abbreviate $\frac{12}{4}$ as 3 (because $\frac{12}{4} = \frac{3}{1}$).

Exercise 1.16: Place the following rational numbers into six groups so that any two rational numbers in the same group are equal, while any two rational numbers in different groups are not equal

$$\frac{3}{5} \quad \frac{-13}{7} \quad \frac{0}{6} \quad \frac{6}{10} \quad \frac{4}{4} \quad 0 \quad \frac{13}{-7} \quad \frac{-5}{-3} \quad \frac{-17}{-17} \quad 1 \quad \frac{1}{1} \quad \frac{10}{-6} \quad \frac{0}{-1} \quad \frac{-15}{9} \quad \frac{20}{12}$$

_____ _____ _____ _____ _____ _____

If a and b are integers and $b \neq 0$, then the expression $\frac{a}{b}$ is called a **fraction**. So, the set of rational numbers, $\mathbb{Q}$, can also be referred to as the set of fractions. Each fraction can also be represented in another way, namely as a **decimal**. We will discuss this a bit more after defining the real numbers.

To define the set of **real numbers**, $\mathbb{R}$, we first define a **digit** to be one of the symbols 0, 1, 2, 3, 4, 5, 6, 7, 8, or 9. We then define $\mathbb{R}$ to be the set of numbers of the form $x.y$ (the dot between the x and y is called a **decimal point** and the number $x.y$ is called a **decimal**), where $x \in \mathbb{Z}$ and y is an infinite "string" of digits without a **tail of 9's** (meaning there are infinitely many digits in the string that are **not** 9). Symbolically, we have

$$\mathbb{R} = \{x.y \mid x \in \mathbb{Z} \text{ and } y \text{ is an infinite string of digits without a tail of 9's}\}.$$

Some examples of real numbers are 0.000 ..., 0.666 ..., -17.000 ..., and 2.020020002 ... We will generally delete tails of 0's. So, we would write 0.000 ... as 0 and -17.000 ... as -17. We will not consider 76.073999999 ... to be a real number because of the tail of 9's (an alternative approach would be to identify 76.073999999 ... with 76.074). We can visualize the set of real numbers with the **real line**.

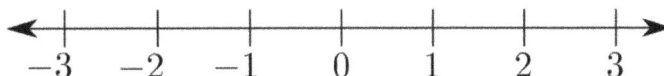

Earlier, we mentioned that we could represent every fraction (rational number) as a decimal. There are two practical ways to do this.

1. Type the fraction into a calculator and press ENTER.

2. Perform long division.

For example, we can represent the fraction $\frac{3}{2}$ as the decimal 1.5 and we can represent the fraction $\frac{2}{3}$ as the real number 0.66666 ... We may abbreviate this last number by using the notation $0.\overline{6}$. The "bar" over the 6 indicates that the 6 repeats forever. As another example of this notation, the number $0.12\overline{345}$ abbreviates 0.12345345345345 ... Notice how the 1 and 2 appear just once (because the bar is not over those digits), whereas the 3, 4, and 5 repeat forever.

Every rational number can be represented as a decimal that either **terminates** (has a tail of 0's) or **repeats** (has a tail with a finite repeating pattern). Any decimal (real number) that does **not** terminate or repeat is called an **irrational number**.

Exercise 1.17: Determine if each real number is a rational number or an irrational number.

1. 6.789 _____

2. 3.626262626 ... _____

3. 1.01001000100001000001 ... _____

4. $-1.23456789123456789123456789$ ... _____

5. 1.234567891011121314151617181 9 ... _____

The **complex numbers** are defined as $\mathbb{C} = \{a + bi \mid a, b \in \mathbb{R}\}$. In words, $\mathbb{C}$ is "the set of $a + bi$ such that a and b are real numbers." Some examples of complex numbers are $0 + 0i$, $-8 + 0i$, $3 + 7i$, $4.5 - 3i = 4.5 + (-3)i$, and $19.235235235\ldots + 84.2020020002\ldots i$. We will abbreviate $0 + 0i$ as 0, $a + 0i$ as a, and $0 + bi$ as bi. For example, $-8 + 0i = -8$. By identifying $a + 0i$ as a, we can think of every real number as a complex number. Complex numbers of the form bi are called **pure imaginary numbers**.

If we identify $1 = 1 + 0i$ with the point $(1, 0)$, and we identify $i = 0 + 1i$ with the point $(0, 1)$, then it is natural to write the complex number $a + bi$ as the point (a, b). Here is a reasonable justification for this: $a + bi = a(1, 0) + b(0, 1) = (a, 0) + (0, b) = (a, b)$

In this way, we can visualize a complex number as a point in **The Complex Plane**. A portion of the Complex Plane is shown to the right with several complex numbers displayed as points of the form (x, y).

The Complex Plane is formed by taking two copies of the real line and placing one horizontally and the other vertically. The horizontal copy of the real line is called the x-axis or the **real axis** (labeled x in the figure) and the vertical copy of the real line is called the y-axis or **imaginary axis** (labeled y in the figure). The two axes intersect at the point $(0, 0)$. This point is called the **origin**.

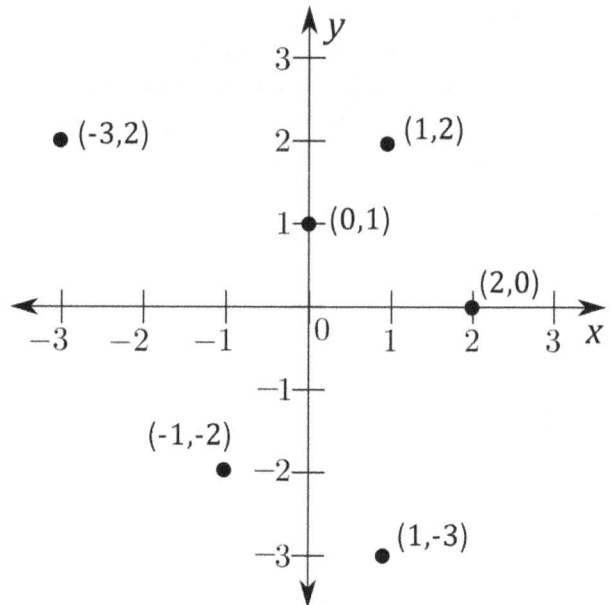

Exercise 1.18: Plot the following points in the Complex Plane:

1. 0
2. -1.5
3. $1 + i$
4. $-2 + \frac{5}{2}i$
5. $-1.\overline{3}i$

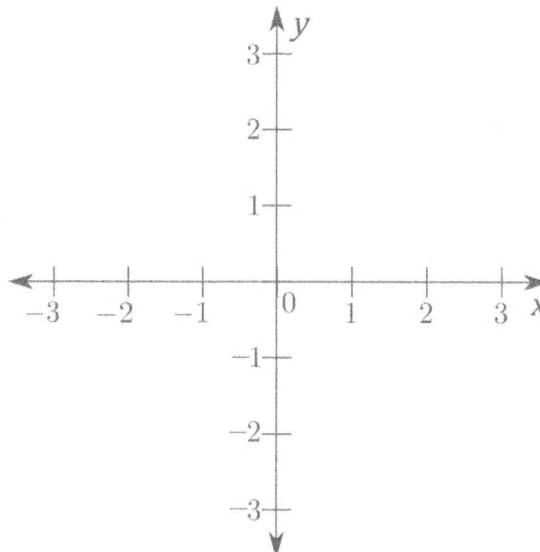

The **empty set** is the unique set with no elements. We use the symbol $\emptyset$ to denote the empty set (some authors use the symbol $\{\,\}$ instead).

Sets Inside of Sets

A set may contain other sets as elements. For example, the set $A = \{a, \{b, c\}\}$ is the set consisting of the two elements a and $\{b, c\}$. It's worth emphasizing that b and c are **not** elements of the set A. Using the membership relation, we have $a \in A$, $\{b, c\} \in A$, $b \notin A$, and $c \notin A$.

Example 1.19:

1. $X = \{\{x\}\}$ is the set consisting of one element: $\{x\}$. Note that $x \notin X$, while $\{x\} \in X$.

2. $Y = \{\{0\}, \{1\}, \{2\}\}$ is the set consisting of three elements: $\{0\}$, $\{1\}$, and $\{2\}$. Note that $0 \notin Y$, $1 \notin Y$, and $2 \notin Y$, while $\{0\} \in Y$, $\{1\} \in Y$, and $\{2\} \in Y$.

3. $Z = \{\emptyset, \{\emptyset\}\}$ is the set consisting of two elements: $\emptyset$ and $\{\emptyset\}$.

Notes: (1) When a set containing other sets is described using the roster method, it is easy to see all the elements by simply removing the outer pair of brackets.

(2) For example, when we remove the outer brackets of $X = \{\{x\}\}$, we get the only element of this set, namely $\{x\}$. Similarly, when we remove the outer brackets of $Y = \{\{0\}, \{1\}, \{2\}\}$, we get the elements $\{0\}, \{1\}, \{2\}$. Let's also remove the outer brackets of $\{\emptyset, \{\emptyset\}\}$. In this case, we get the elements $\emptyset, \{\emptyset\}$.

(3) There is actually no requirement that the elements of the set under consideration be sets for this technique to work. For example, if $A = \{a, b, c, d\}$, then we can still see all the elements by removing the outer brackets. Indeed, the elements of A are $a, b, c,$ and d.

Exercise 1.20: Determine how many elements are in each of the following sets and then list the elements in the set.

1. $\{\{a\}, \{b, c\}, \{d, e, f\}\}$ _____

2. $\left\{\{x\}, \{x, \{x\}\}, \{\{x\}\}, \{x, \{x\}, \{\{x\}\}\}\right\}$ _____

3. $\left\{\{\emptyset, \{\emptyset\}, \{\emptyset, \{\emptyset\}\}\}\right\}$ _____

4. $\{\{z\}, \{z, z\}, \{z, z, z\}\}$ _____

Cardinality of a Finite Set

If A is a finite set, we define the **cardinality** of A, written $|A|$, to be the number of elements of A. For example, $|\{0, 1, 2, 3, 4\}| = 5$.

Example 1.21: Let $A = \{\text{rat}, \text{squirrel}, \text{ruby}, \text{chair}\}$, $B = \{c, d, c\}$, $C = \{25, 26, 27, \ldots, 3167, 3168\}$, $D = \{\{a\}, \{a\}, \{a, a\}, \{a, a, a\}\}$, and $E = \emptyset$. Then $|A| = 4$, $|B| = 2$, $|C| = 3144$, $|D| = 1$, and $|E| = 0$.

Notes: (1) The set A consists of the four elements "rat," "squirrel," "ruby," and "chair."

(2) The set B consists of just two elements: c and d. Remember that $\{c, d, c\} = \{c, d\}$.

(3) The number of consecutive integers from m to n, inclusive, is $n - m + 1$. For set C, we have $m = 25$ and $n = 3168$. Therefore, $|C| = 3168 - 25 + 1 = 3144$.

(4) I call the formula "$n - m + 1$" the **fence-post formula**. If you construct a 3-foot fence by placing a fence-post every foot, then the fence will consist of 4 fence-posts ($3 - 0 + 1 = 4$).

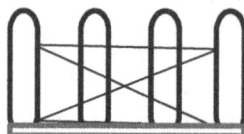

(5) Since $\{a, a\} = \{a\}$ and $\{a, a, a\} = \{a\}$, it follows that $D = \{\{a\}, \{a\}, \{a\}, \{a\}\} = \{\{a\}\}$. So, D consists of the single element $\{a\}$.

(6) Remember that $\emptyset$ (pronounced "the empty set") is the unique set with no elements.

Exercise 1.22: Determine the cardinality of each of the following sets:

1. $\{1, 2, 3, \ldots, 80\}$ _____

2. $\{c, d, e, f, e, d, c\}$ _____

3. $\left\{\emptyset, \{\emptyset\}, \{\emptyset, \{\emptyset\}\}\right\}$ _____

4. $\{n \in \mathbb{N} \mid 222 \leq n \leq 2222\}$ _____

5. $\left\{x, \{x, x\}, \{x, x, x\}, \{x, \{x, x\}\}\right\}$ _____

Subsets and Proper Subsets

We say that a set A is a **subset** of a set B, written $A \subseteq B$, if every element of A is an element of B.

Example 1.23:

1. Let $A = \{x, y\}$ and $B = \{x, y, z\}$. The only elements of A are x and y. Since x and y are also elements of B, we see that $A \subseteq B$.

   Notice that $B \nsubseteq A$ (B is **not** a subset of A) because $z \in B$, but $z \notin A$.

2. Let $\mathbb{N} = \{0, 1, 2, 3, \ldots\}$ be the set of natural numbers and let $\mathbb{Z} = \{\ldots, -3, -2, -1, 0, 1, 2, 3, 4, \ldots\}$ be the set of integers. Since every natural number is an integer, $\mathbb{N} \subseteq \mathbb{Z}$.

3. By making appropriate identifications, we have the following sequence of inclusions:

$$\mathbb{N} \subseteq \mathbb{Z} \subseteq \mathbb{Q} \subseteq \mathbb{R} \subseteq \mathbb{C}.$$

   In general, if $A \subseteq B$ and $B \subseteq C$, then $A \subseteq C$ (we say that $\subseteq$ is **transitive**). In this way we see that we have many other inclusions such as $\mathbb{N} \subseteq \mathbb{Q}$, $\mathbb{N} \subseteq \mathbb{R}, \ldots$, and so on.

4. Consider the sets $2\mathbb{Z} = \{\ldots, -6, -4, -2, 0, 2, 4, 6, \ldots\}$ and $4\mathbb{Z} = \{\ldots, -12, -8, -4, 0, 4, 8, 12, \ldots\}$. Then $4\mathbb{Z} \subseteq 2\mathbb{Z}$. Note that the opposite inclusion is false. That is, $2\mathbb{Z} \nsubseteq 4\mathbb{Z}$. To see this, we just need a single **counterexample** (a counterexample is an example that is used to show that a statement is false). Well, we have $2 \in 2\mathbb{Z}$, but $2 \notin 4\mathbb{Z}$.

5. Consider the sets $2\mathbb{Z} = \{\ldots, -6, -4, -2, 0, 2, 4, 6, \ldots\}$ and $3\mathbb{Z} = \{\ldots, -9, -6, -3, 0, 3, 6, 9, \ldots\}$. Neither of these sets is a subset of the other. To see that $2\mathbb{Z} \nsubseteq 3\mathbb{Z}$, observe that $2 \in 2\mathbb{Z}$, whereas $2 \notin 3\mathbb{Z}$. To see that $3\mathbb{Z} \nsubseteq 2\mathbb{Z}$, observe that $3 \in 3\mathbb{Z}$, whereas $3 \notin 2\mathbb{Z}$.

6. Let $A = \{a\}$ and $B = \{a, a\}$. As we already know, these two sets are equal (listing an element of a set more than once is equivalent to listing that element just once). When two sets are equal, they are subsets of each other. That is, if $X = Y$, then $X \subseteq Y$ and $Y \subseteq X$ (we can abbreviate this as "$X = Y \to (X \subseteq Y \wedge Y \subseteq X)$"). Conversely, if $X \subseteq Y$ and $Y \subseteq X$, then $X = Y$ (we can abbreviate this as "$(X \subseteq Y \wedge Y \subseteq X) \to X = Y$").

   We can combine the previous two statements into the single statement $X = Y$ **if and only if** $X \subseteq Y$ and $Y \subseteq X$ (which we can abbreviate as "$X = Y \leftrightarrow (X \subseteq Y \wedge Y \subseteq X)$"). We can think of the expression "if and only if" as two "if...then" statements. Another way to say this is that the statements "$X = Y$" and "$X \subseteq Y$ and $Y \subseteq X$" are equivalent.

   The statement "$X = Y$ if and only if $X \subseteq Y$ and $Y \subseteq X$" is known as the **Axiom of Extensionality**. An **axiom** is simply a statement that is assumed to be true. See Lesson 7 for more about this axiom and the other axioms of set theory.

Exercise 1.24: For each pair of sets A and B below, determine if $A \subseteq B$, $B \subseteq A$, both, or neither.

1. $A = \{a, b, c\}, B = \{a, c\}$ _____

2. $A = 2\mathbb{Z}, B = \mathbb{N}$ _____

3. $A = \{x, \{x, x\}\}, B = \{\{x\}, x, x\}$ _____

4. $A = 3\mathbb{Z}, B = \{t \in \mathbb{Z} \mid t = 0, 3, 6, 9, \ldots\}$ _____

To the right we see a physical representation of $A \subseteq B$. This figure is called a **Venn diagram**. These types of diagrams are very useful to help visualize relationships among sets. Notice that set A lies completely inside set B. We assume that all the elements of A and B lie in some **universal set** U.

As an example, let U be the set of all species of animals. If we let A be the set of species of cats and we let B be the set of species of mammals, then we have $A \subseteq B \subseteq U$, and we see that the Venn diagram to the right gives a visual representation of this situation. (Note that every cat is a mammal and every mammal is an animal.)

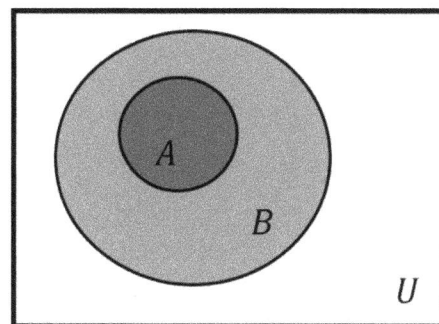

$A \subseteq B$

We say that A is a **proper subset** of B, written $A \subset B$ (or sometimes $A \subsetneq B$), if $A \subseteq B$, but $A \neq B$. For example, $\mathbb{N} \subset \mathbb{Z}$, whereas $\mathbb{N} \not\subset \mathbb{N}$ (although $\mathbb{N} \subseteq \mathbb{N}$).

Note: The definition of proper subset is not too important. It just gives us a convenient way to discuss all the subsets of a specific set except the set itself. For example, it is quite cumbersome to say "Find all subsets of A, but exclude the set A." Instead, we can rephrase this as "Find all proper subsets of A."

The following basic facts about subsets are useful.

Subset Fact 1: Every set is a subset of itself.

Subset Fact 2: The empty set is a subset of every set.

Subset Fact 3: $\subseteq$ is **transitive**. In other words, if $A \subseteq B$ and $B \subseteq C$, then $A \subseteq C$.

Note: To the right we have a Venn diagram illustrating the transitivity of $\subseteq$ (Subset Fact 3).

Since $\subseteq$ is transitive, we can write things like $A \subseteq B \subseteq C \subseteq D$, and without explicitly saying it, we know that $A \subseteq C$, $A \subseteq D$, and $B \subseteq D$.

For example, in part 3 of Example 1.23 above, we saw that

$$\mathbb{N} \subseteq \mathbb{Z} \subseteq \mathbb{Q} \subseteq \mathbb{R} \subseteq \mathbb{C}.$$

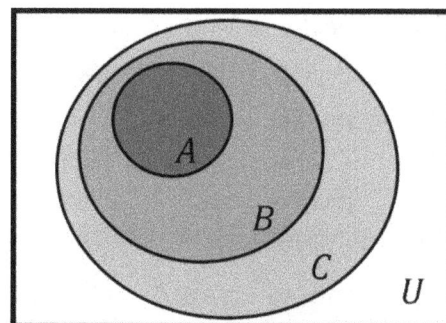

$A \subseteq B \subseteq C$

Since $\subseteq$ is transitive, we automatically know that $\mathbb{N} \subseteq \mathbb{Q}$, $\mathbb{N} \subseteq \mathbb{R}$, $\mathbb{N} \subseteq \mathbb{C}$, $\mathbb{Z} \subseteq \mathbb{R}$, $\mathbb{Z} \subseteq \mathbb{C}$, and $\mathbb{Q} \subseteq \mathbb{C}$.

Example 1.25: Let $C = \{a, b, c\}$, $D = \{a, c\}$, $E = \{b, c\}$, $F = \{b, d\}$, and $G = \emptyset$. Then $D \subseteq C$ and $E \subseteq C$. Also, since **the empty set is a subset of every set**, we have $G \subseteq C$, $G \subseteq D$, $G \subseteq E$, $G \subseteq F$, and $G \subseteq G$. **Every set is a subset of itself**, and so, $C \subseteq C$, $D \subseteq D$, $E \subseteq E$, and $F \subseteq F$.

Exercise 1.26: Draw a Venn Diagram displaying the sets C, D, E, and F from Example 1.25 inside a universal set U.

Power Sets

If A is a set, then the **power set** of A, written $\mathcal{P}(A)$, is the set of all subsets of A. In set-builder notation, we write $\mathcal{P}(A) = \{B \mid B \subseteq A\}$.

Example 1.27: The set $A = \{a, b\}$ has 2 elements and 4 subsets. The subsets of A are $\emptyset$, $\{a\}$, $\{b\}$, and $\{a, b\}$. It follows that $\mathcal{P}(A) = \{\emptyset, \{a\}, \{b\}, \{a, b\}\}$.

The set $B = \{a, b, c\}$ has 3 elements and 8 subsets. The subsets of B are $\emptyset$, $\{a\}$, $\{b\}$, $\{c\}$, $\{a, b\}$, $\{a, c\}$, $\{b, c\}$, and $\{a, b, c\}$. It follows that $\mathcal{P}(B) = \{\emptyset, \{a\}, \{b\}, \{c\}, \{a, b\}, \{a, c\}, \{b, c\}, \{a, b, c\}\}$.

Let's draw a **tree diagram** for the subsets of each of the sets A and B.

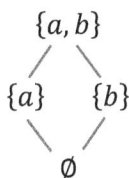

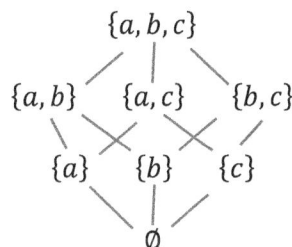

The tree diagram above on the left is for the subsets of the set $A = \{a, b\}$. We start by writing the set $A = \{a, b\}$ at the top. On the next line we write the subsets of cardinality 1 ($\{a\}$ and $\{b\}$). On the line below that we write the subsets of cardinality 0 (just $\emptyset$). We draw a line segment between any two sets when the smaller (lower) set is a subset of the larger (higher) set. So, we see that $\emptyset \subseteq \{a\}$, $\emptyset \subseteq \{b\}$, $\{a\} \subseteq \{a, b\}$, and $\{b\} \subseteq \{a, b\}$. There is actually one more subset relationship, namely $\emptyset \subseteq \{a, b\}$ (and of course each set displayed is a subset of itself). We didn't draw a line segment from $\emptyset$ to $\{a, b\}$ to avoid unnecessary clutter. Instead, we can simply trace the path from $\emptyset$ to $\{a\}$ to $\{a, b\}$ (or from $\emptyset$ to $\{b\}$ to $\{a, b\}$). We are using the **transitivity** of $\subseteq$ here.

The tree diagram above on the right is for the subsets of $B = \{a, b, c\}$. Observe that from top to bottom we write the subsets of B of cardinality 3, then 2, then 1, and then 0. We then draw the appropriate line segments, just as we did for $A = \{a, b\}$.

Exercise 1.28: How many subsets does $\{a, b, c, d\}$ have? _____ Draw a tree diagram for the subsets of $\{a, b, c, d\}$.

Example 1.29:

1. A set with 0 elements must be $\emptyset$, and this set has exactly 1 subset (the only subset of the empty set is the empty set itself).

2. A set with 1 element has 2 subsets, namely $\emptyset$ and the set itself.

3. In Example 1.27, we saw that a set with 2 elements has 4 subsets, and we saw that a set with 3 elements has 8 subsets.

4. How many subsets does a set of cardinality n have? Do you see the pattern from parts 1, 2, and 3 above? Well, $1 = 2^0$, $2 = 2^1$, $4 = 2^2$, $8 = 2^3$. So, we see that a set with 0 elements has 2^0 subsets, a set with 1 element has 2^1 subsets, a set with 2 elements has 2^2 subsets, and a set with 3 elements has 2^3 subsets.

   It seems that in general, a set with n elements has $\mathbf{2^n}$ subsets. We can also say that if $|A| = n$, then $|\mathcal{P}(A)| = 2^n$.

Exercise 1.30: Let X be a set such that $|\mathcal{P}(X)| = 128$. What is $|X|$? ____

The Natural Numbers

At this point, let's provide more formal definitions of the natural numbers and the set of natural numbers.

We define the following:

$$0 = \emptyset$$
$$1 = \{\emptyset\} = \{0\}$$
$$2 = \{\emptyset, \{\emptyset\}\} = \{0, 1\}$$
$$3 = \{\emptyset, \{\emptyset\}, \{\emptyset, \{\emptyset\}\}\} = \{0, 1, 2\}$$
$$\ldots \qquad \ldots$$

In general, we let $n = \{0, 1, 2, \ldots, n - 1\}$.

We now define the set of **natural numbers** to be $\mathbb{N} = \{0, 1, 2, \ldots\}$.

Exercise 1.31: Explicitly write down the natural number 4 using only set brackets and the empty set.

___

Problem Set 1

Full solutions to these problems are available for free download here:
www.SATPrepGet800.com/STKZ3D

LEVEL 1

Determine whether each of the following statements is true or false:

1. $t \in \{t\}$

2. $4 \in \{0, 2, 4, 6\}$

3. $-7 \in \{7\}$

4. $0 \in \mathbb{Z}$

5. $-26 \in \mathbb{N}$

6. $\frac{7}{2} \in \mathbb{Q}$

7. $\emptyset \subseteq \{1, 2, 3\}$

8. $\{\square\} \subseteq \{\square, \Delta\}$

9. $\{x, y, z, w\} \subset \{x, y, z, w\}$

10. $\{0, 1, \{2, 3\}\} \subseteq \{0, 1, 2, 3\}$

Determine the cardinality of each of the following sets:

11. $\{\text{hammer}, \text{screwdriver}, \text{saw}\}$

12. $\{4, 16, 27, 46, 59, 201\}$

13. $\{1, 2, \dots, 236\}$

14. $\left\{\frac{1}{2}, \frac{1}{3}, \dots, \frac{1}{15}\right\}$

15. $\emptyset$

Provide an example of a set X with the given properties:

16. (i) $X \subset \mathbb{Z}$ (X is a *proper* subset of $\mathbb{Z}$); (ii) X is infinite; (iii) X contains both positive and negative integers; (iv) X contains both even and odd integers.

17. (i) $X \subset \mathbb{R}$ (X is a *proper* subset of $\mathbb{R}$); (ii) X contains both rational and irrational numbers.

18. (i) $X \subset \mathbb{C}$ (X is a *proper* subset of $\mathbb{C}$); (ii) X is infinite; (iii) X contains real numbers; and (iv) X contains complex numbers that are not real.

For each pair of sets A and B below, determine if $A \subseteq B$, $B \subseteq A$, both, or neither.

19. $A = 2\mathbb{N}$, $B = \mathbb{Z}$

20. $A = \{0, 1\}$, $B = \{1, 0\}$

21. $A = \emptyset$, $B = \{\}$

22. $A = \{0, 1, 2, 3, 4\}$, $B = \{1, 3\}$

23. $A = \{a\}$, $B = \{b\}$

LEVEL 2

Determine whether each of the following statements is true or false:

24. $a \in \emptyset$

25. $\emptyset \in \emptyset$

26. $\emptyset \in \{\emptyset, \{\emptyset\}\}$

27. $\{\emptyset\} \in \emptyset$

28. $\{\emptyset\} \in \{\emptyset\}$

29. $7 \in 2\mathbb{N}$

30. $\emptyset \subseteq \emptyset$

31. $\emptyset \subseteq \{\emptyset\}$

32. $\{\emptyset\} \subseteq \emptyset$

33. $\{\emptyset\} \subseteq \{\emptyset\}$

Determine the cardinality of each of the following sets:

34. $\{a, a, b, c, c, c, d, d\}$

35. $\{\{x, y\}, \{z, u, v\}, \{w\}\}$

36. $\{5, 6, 7, \dots, 4322, 4323\}$

Determine if each of the following real numbers is rational or irrational:

37. $5.\overline{7}$

38. $-72.810121416182022242628\dots$

39. $463.65432154321543215432154321\dots$

40. 0

LEVEL 3

Use set-builder notation to describe each of the following sets.

41. $\{2, 4, 6, 8, 10, 12, 14, 16, 18, 20, 22\}$

42. $2\mathbb{N}$

43. $\mathbb{Z}^-$

Determine the cardinality of each of the following sets:

44. $\{\{\{x, y, z\}\}\}$

45. $\{\{0, 1\}, 0, \{0\}, \{0, \{0, 1, 2\}\}\}$

46. $\{a, \{a\}, \{a, a\}, \{a, a, a, a\}, \{a, a, \{a\}\}, \{a, \{a\}, \{a\}\}\}$

For each set X, determine $|\mathcal{P}(X)|$

    47. $X = \{0, 1, 2, 3, 4, 5, 6\}$

    48. $X = \{\emptyset, \{\emptyset\}, \{\emptyset, \{\emptyset\}\}\}$

    49. $X = \{26, 27, 28, \dots, 203, 204\}$

LEVEL 4

Determine whether each of the following statements is true or false:

    50. $x \in \{x, \{y\}\}$

    51. $3 \in \{2k \mid k = 1, 2, 3, 4\}$

    52. $\{0\} \in \{0, 1\}$

    53. $\{1\} \in \{\{1\}, x, 2, y\}$

    54. $\emptyset \in \{\{\emptyset\}\}$

    55. $\{\{\emptyset\}\} \in \emptyset$

Compute the power set of each of the following sets:

    56. $\emptyset$

    57. $\{\emptyset\}$

    58. $\{lion, tiger, jaguar\}$

    59. $\{\emptyset, \{\emptyset\}\}$

    60. $\{\{\emptyset\}\}$

    61. $\{\emptyset, \{\emptyset\}, \{\emptyset, \{\emptyset\}\}\}$

A **relation** describes a relationship between objects. For example, the relation $=$ describes the relationship "is equal to." Two other relations we have seen are $\in$ (the membership relation) and $\subseteq$ (the subset relation). A relation R is **reflexive** if for all x, we have xRx. A relation R is **symmetric** if for all x,y, we have $xRy \rightarrow yRx$. A relation R is **transitive** if for all x,y, z, we have $(xRy \land yRz) \rightarrow xRz$. For example, the relation "$=$" is reflexive, symmetric, and transitive because for all x, we have $x = x$, for all x,y, we have $x = y \rightarrow y = x$, and for all x,y, z, we have $(x = y \land y = z) \rightarrow x = z$.

62. Is $\subseteq$ reflexive?

63. Is $\in$ reflexive?

64. Is $\subseteq$ symmetric?

65. Is $\in$ symmetric?

66. Is $\subseteq$ transitive?

67. Is $\in$ transitive?

Explicitly write down each of the following natural numbers using only set brackets and the empty set.

68. 5

69. 6

LEVEL 5

We say that a set A is **transitive** if every element of A is a subset of A. Determine if each of the following sets is transitive:

70. $\emptyset$

71. $\{\emptyset\}$

72. $\{\{\emptyset\}\}$

73. $\{\emptyset, \{\emptyset\}\}$

74. $\{\emptyset, \{\emptyset\}, \{\{\emptyset\}\}\}$

75. $\{\{\emptyset\}, \{\emptyset, \{\emptyset\}\}\}$

76. Assuming that A is transitive, is $\mathcal{P}(A)$ transitive?

Let A and B be sets with $B \subseteq A$. Determine if the following are true or false.

77. $B \in \mathcal{P}(A)$

78. $B \subseteq \mathcal{P}(A)$

79. $\mathcal{P}(B) \in \mathcal{P}(A)$

80. $\mathcal{P}(B) \subseteq \mathcal{P}(A)$

CHALLENGE PROBLEMS

81. Let $A = \{a, b, c, d\}$, $B = \{X \mid X \subseteq A \wedge d \notin X\}$, and $C = \{X \mid X \subseteq A \wedge d \in X\}$. Show that there is a natural **one-to-one correspondence** (see definition below) between the elements of B and the elements of C. Then generalize this result to a set with $n + 1$ elements for $n > 0$.

    Definition: A **one-to-one correspondence** between two sets is a pairing so that each element of the first set is matched up with exactly one element of the second set, and vice versa.

82. Recall that a set A is **transitive** if every element of A is a subset of A. Provide an example of an infinite transitive set

83. Let A and B be sets with $A \subseteq B$ and B transitive. Determine if $\mathcal{P}(A) \subseteq \mathcal{P}\big(\mathcal{P}(B)\big)$.

LESSON 2
OPERATIONS ON SETS

Basic Set Operations

The **union** of the sets A and B, written $A \cup B$, is the set of elements that are in A or B (or both).

$$A \cup B = \{x \mid x \in A \text{ or } x \in B\}$$

The **intersection** of A and B, written $A \cap B$, is the set of elements that are simultaneously in A and B.

$$A \cap B = \{x \mid x \in A \text{ and } x \in B\}$$

The following Venn diagrams for the union and intersection of two sets can be useful for visualizing these operations.

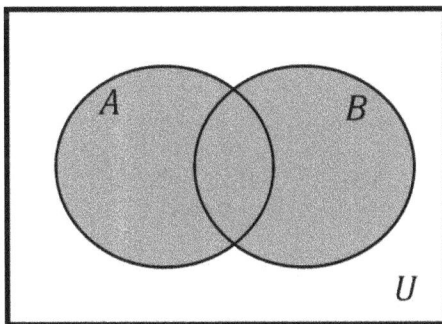

$A \cup B$

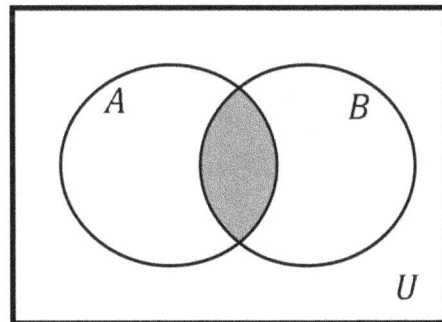

$A \cap B$

The **difference** $A \setminus B$ is the set of elements that are in A and not in B.

$$A \setminus B = \{x \mid x \in A \text{ and } x \notin B\}$$

The **symmetric difference** between A and B, written $A \, \Delta \, B$, is the set of elements that are in A or B, but not both.

$$A \, \Delta \, B = (A \setminus B) \cup (B \setminus A)$$

Let's also look at Venn diagrams for the difference and symmetric difference of two sets.

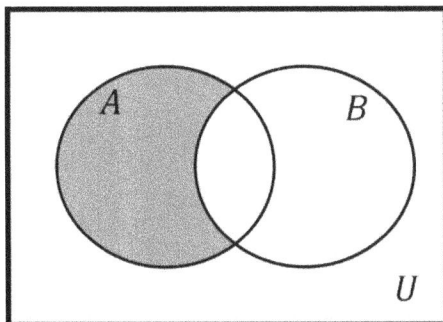

$A \setminus B$

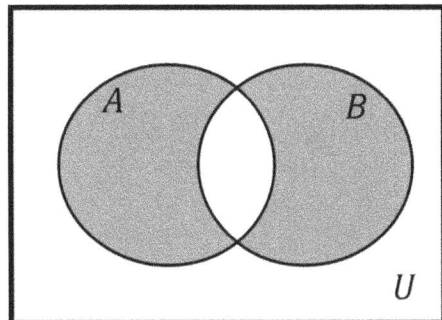

$A \, \Delta \, B$

Example 2.1: Let $A = \{0, 1, 2, 3, 4\}$ and $B = \{3, 4, 5, 6\}$. We have

1. $A \cup B = \{0, 1, 2, 3, 4, 5, 6\}$.
2. $A \cap B = \{3, 4\}$.
3. $A \setminus B = \{0, 1, 2\}$.
4. $B \setminus A = \{5, 6\}$.
5. $A \Delta B = \{0, 1, 2\} \cup \{5, 6\} = \{0, 1, 2, 5, 6\}$.

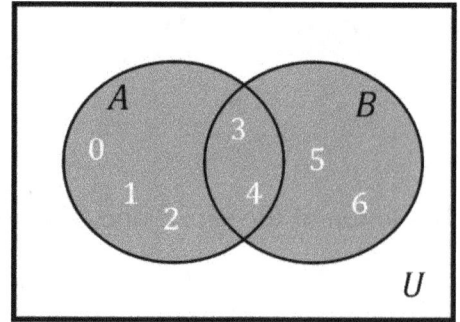

Exercise 2.2: Let $A = \{a, b, \Delta, \delta\}$ and $B = \{b, c, \delta, \gamma\}$. Determine each of the following:

1. $A \cup B$ _____
2. $A \cap B$ _____
3. $A \setminus B$ _____
4. $B \setminus A$ _____
5. $A \Delta B$ _____

Example 2.3: Recall that the set of natural numbers is $\mathbb{N} = \{0, 1, 2, 3, \ldots\}$ and the set of integers is $\mathbb{Z} = \{\ldots, -4, -3, -2, -1, 0, 1, 2, 3, 4, \ldots\}$. Observe that in this case, $\mathbb{N} \subseteq \mathbb{Z}$. We have

1. $\mathbb{N} \cup \mathbb{Z} = \mathbb{Z}$.
2. $\mathbb{N} \cap \mathbb{Z} = \mathbb{N}$.
3. $\mathbb{N} \setminus \mathbb{Z} = \emptyset$.
4. $\mathbb{Z} \setminus \mathbb{N} = \{\ldots, -4, -3, -2, -1\} = \mathbb{Z}^-$. (Recall that $\mathbb{Z}^-$ is "the set of negative integers.")
5. $\mathbb{N} \Delta \mathbb{Z} = \emptyset \cup \mathbb{Z}^- = \mathbb{Z}^-$.

Note: Whenever A and B are sets and $B \subseteq A$, then

1. $A \cup B = A$.
2. $A \cap B = B$.
3. $B \setminus A = \emptyset$.

Example 2.4: Let $\mathbb{E} = 2\mathbb{N} = \{0, 2, 4, 6, \ldots\}$ be the set of even natural numbers and let $\mathbb{O} = 2\mathbb{N} + 1 = \{1, 3, 5, 7, \ldots\}$ be the set of odd natural numbers. We have

1. $\mathbb{E} \cup \mathbb{O} = \{0, 1, 2, 3, 4, 5, 6, 7, \ldots\} = \mathbb{N}$.
2. $\mathbb{E} \cap \mathbb{O} = \emptyset$.
3. $\mathbb{E} \setminus \mathbb{O} = \mathbb{E}$.
4. $\mathbb{O} \setminus \mathbb{E} = \mathbb{O}$.
5. $\mathbb{E} \Delta \mathbb{O} = \mathbb{E} \cup \mathbb{O} = \mathbb{N}$.

In general, we say that sets A and B are **disjoint** or **mutually exclusive** if $A \cap B = \emptyset$. To the right is a Venn diagram for disjoint sets.

In Example 2.4 above, we saw that the sets $\mathbb{E} = 2\mathbb{N}$ and $\mathbb{O} = 2\mathbb{N} + 1$ are disjoint.

Exercise 2.5: Consider the sets $A = \{a + bi \in \mathbb{C} \mid a, b \in \mathbb{Z}\}$ and $B = \{a + bi \in \mathbb{C} \mid a \notin \mathbb{Q}\}$.

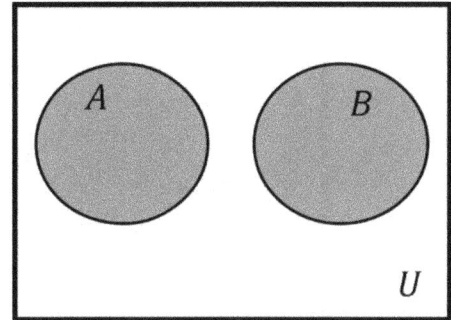

$A \cap B = \emptyset$

1. Is $A \subseteq B$? _____

2. Is $B \subseteq A$? _____

3. Are A and B disjoint? _____

The following basic facts about unions, intersections, and set differences are useful (some of these were mentioned in the Note following Example 2.3).

Set Operation Fact 1: $A \subseteq A \cup B$.

Set Operation Fact 2: $A \cap B \subseteq A$.

Set Operation Fact 3: $B \subseteq A$ if and only if $A \cup B = A$.

Set Operation Fact 4: $B \subseteq A$ if and only if $A \cap B = B$.

Set Operation Fact 5: $B \subseteq A$ if and only if $B \setminus A = \emptyset$.

Exercise 2.6: Show that each of the following statements is false by providing a counterexample.

1. $A \cup B \subseteq A$. $A =$_____ $B =$_____

2. $A \subseteq A \cap B$. $A =$_____ $B =$_____

3. If $B \subseteq A$, then $A \cup B = B$. $A =$_____ $B =$_____

4. If $B \subseteq A$, then $A \setminus B = \emptyset$. $A =$_____ $B =$_____

Unions, intersections, and set differences have many nice algebraic properties such as the following:

1. **Commutativity:** $A \cup B = B \cup A$ and $A \cap B = B \cap A$.

2. **Associativity:** $(A \cup B) \cup C = A \cup (B \cup C)$ and $(A \cap B) \cap C = A \cap (B \cap C)$.

3. **Distributivity:** $A \cap (B \cup C) = (A \cap B) \cup (A \cap C)$ and $A \cup (B \cap C) = (A \cup B) \cap (A \cup C)$.

4. **De Morgan's Laws:** $C \setminus (A \cup B) = (C \setminus A) \cap (C \setminus B)$ and $C \setminus (A \cap B) = (C \setminus A) \cup (C \setminus B)$.

5. **Idempotent Laws:** $A \cup A = A$ and $A \cap A = A$.

Example 2.7: As an example, we can use the following Venn Diagrams to see why the operation of forming unions is associative.

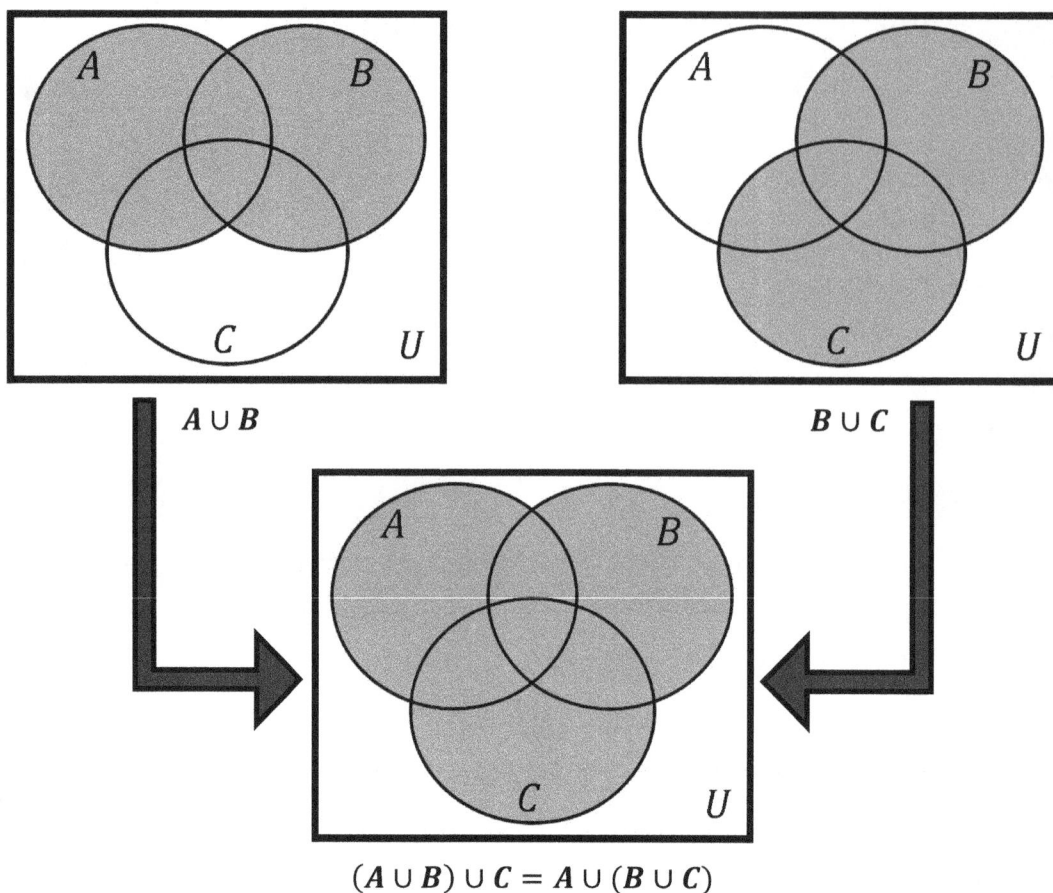

$$(A \cup B) \cup C = A \cup (B \cup C)$$

The dedicated reader may want to draw Venn Diagrams to help visualize the rest of the properties.

Remember that associativity allows us to drop parentheses. So, we can now simply write $A \cup B \cup C$ when taking the union of the three sets A, B, and C.

Intervals of Real Numbers

A set I of real numbers with at least two elements is called an **interval** if any real number that lies between two numbers in I is also in I.

Example 2.8:

1. The set $A = \{0, 1\}$ is **not** an interval. A consists of just the two real numbers 0 and 1. There are infinitely many real numbers between 0 and 1. For example, the real number $\frac{1}{2}$ is between 0 and 1.

2. The set $B = \{x \in \mathbb{R} \mid 0 < x < 1\}$ is an example of an **open interval**. This set consists of all real numbers between 0 and 1, exclusive (0 and 1 are excluded). We will usually write the set B using the **interval notation** $B = (0, 1)$.

3. The set $C = \{x \in \mathbb{R} \mid 0 \leq x \leq 1\}$ is an example of a **closed interval**. This set consists of all real numbers between 0 and 1, inclusive (0 and 1 are included). We will usually write the set C using the **interval notation** $C = [0, 1]$.

4. The set $D = \{x \in \mathbb{R} \mid 0 \le x < 1\}$ is an example of a **half-open interval**. This set consists of all real numbers between 0 and 1, including 0, but excluding 1. We will usually write the set D using the **interval notation** $D = [0, 1)$.

5. The set $E = \{x \in \mathbb{R} \mid x > 1\}$ is an example of an **infinite open interval**. This set consists of all real numbers greater than 1. We will usually write the set E using the **interval notation** $E = (1, \infty)$. The symbol "∞" is pronounced "**infinity**." It is **not** a number, but rather a symbol indicating that the set E has no upper bound.

6. $\mathbb{R}$ is an interval. This follows trivially from the definition. After all, any real number that lies between two real numbers is a real number.

Note: Observe how the "interval notation" mentioned in 2, 3, 4, and 5 above uses parentheses "( )" to indicate that endpoints are **not** included in the set and brackets "[ ]" to indicate that endpoints are included in the set.

Exercise 2.9: Determine if each of the following sets of real numbers is an interval.

1. $F = \{x \in \mathbb{R} \mid 0 < x \le 1\}$ _____

2. $\mathbb{Z}$ _____

3. $G = \left\{ \frac{1}{n} \,\middle|\, n \in \mathbb{Z}^+ \right\}$ _____

4. $H = \left\{ x \in \mathbb{R} \,\middle|\, \frac{1}{5} < x < \frac{1}{4} \right\}$ _____

5. $I = \{x \in \mathbb{R} \mid x \le 2.7\}$ _____

6. $\mathbb{Q}$ _____

When we are thinking of $\mathbb{R}$ as an interval, we sometimes use the notation $(-\infty, \infty)$ and refer to this as **the real line**. The following picture gives the standard geometric interpretation of the real line.

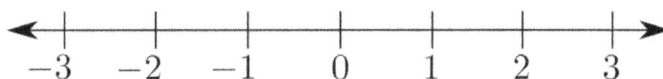

In addition to the real line, there are 8 other types of intervals.

Open Interval:	$(a, b) = \{x \in \mathbb{R} \mid a < x < b\}$
Closed Interval:	$[a, b] = \{x \in \mathbb{R} \mid a \le x \le b\}$

Half-open Intervals: $\quad (a, b] = \{x \in \mathbb{R} \mid a < x \le b\}$ $\qquad [a, b) = \{x \in \mathbb{R} \mid a \le x < b\}$

Infinite Open Intervals: $\quad (a, \infty) = \{x \in \mathbb{R} \mid x > a\}$ $\qquad (-\infty, b) = \{x \in \mathbb{R} \mid x < b\}$

Infinite Closed Intervals: $\quad [a, \infty) = \{x \in \mathbb{R} \mid x \ge a\}$ $\qquad (-\infty, b] = \{x \in \mathbb{R} \mid x \le b\}$

Notes: (1) Each of the nine types of sets above (including the real line) satisfies the definition of being an interval. Conversely, every interval has one of these nine forms.

(2) The first four intervals above (the open, closed, and two half-open intervals) are **bounded**. They are each **bounded below** by a and **bounded above** by b. The last four intervals (the infinite open and closed intervals) are **unbounded**.

31

Example 2.10: The half-open interval $(-2, 1] = \{x \in \mathbb{R} \mid -2 < x \leq 1\}$ has the following graph:

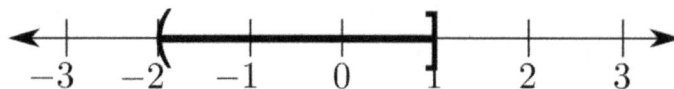

Note: The left parenthesis appearing at -2 indicates that -2 is **not** included in the set, whereas the right bracket appearing at 1 indicates that 1 is included in the set.

Exercise 2.11: Sketch the graph of the infinite open interval $(0, \infty) = \{x \in \mathbb{R} \mid x > 0\}$.

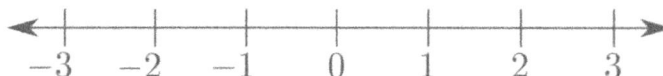

Example 2.12: Let $A = (-2, 1]$ and $B = (0, \infty)$. We have

1. $A \cup B = (-2, \infty)$
2. $A \cap B = (0, 1]$
3. $A \setminus B = (-2, 0]$
4. $B \setminus A = (1, \infty)$
5. $A \mathbin{\Delta} B = (-2, 0] \cup (1, \infty)$

Note: If you have trouble seeing how to compute these, it may be helpful to draw the graphs of A and B lined up vertically, and then draw vertical lines through the endpoints of each interval.

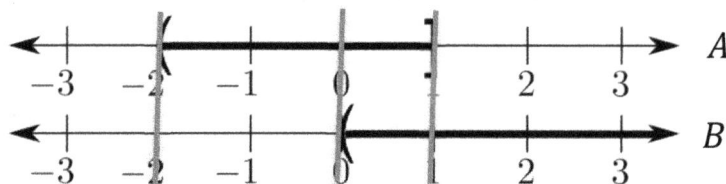

The results follow easily by combining these graphs into a single graph using the vertical lines as guides. For example, let's look at $A \cap B$ in detail. We're looking for all numbers that are in both A and B. The two rightmost vertical lines drawn passing through the two graphs above isolate all those numbers nicely. We see that all numbers between 0 and 1 are in the intersection. We should then think about the two endpoints 0 and 1 separately. $0 \notin B$ and therefore, 0 cannot be in the intersection of A and B. On the other hand, $1 \in A$ and $1 \in B$. Therefore, $1 \in A \cap B$. So, we see that $A \cap B = (0, 1]$.

Exercise 2.13: Let $C = (-\infty, 2]$, $D = (-1, 3]$. Compute each of the following:

1. $C \cup D$ _____
2. $C \cap D$ _____
3. $C \setminus D$ _____
4. $D \setminus C$ _____
5. $C \mathbin{\Delta} D$ _____

32

More Set Operations

We will often be interested in taking unions and intersections of more than two sets. Therefore, we make the following more general definitions.

Let X be a nonempty set of sets.

$$\cup X = \{y \mid \text{there is } Y \in X \text{ with } y \in Y\} \qquad \text{and} \qquad \cap X = \{y \mid \text{for all } Y \in X, y \in Y\}.$$

If you're having trouble understanding what these definitions are saying, you're not alone. The notation probably looks confusing, but the ideas behind these definitions are very simple. You have a whole bunch of sets (possibly infinitely many). To take the union of all these sets, you simply throw all the elements together into one big set. To take the intersection of all these sets, you take only the elements that are in every single one of those sets.

Example 2.14:

1. Let A and B be sets and let $X = \{A, B\}$. Then

    $$\cup X = \{y \mid \text{there is } Y \in X \text{ with } y \in Y\} = \{y \mid y \in A \text{ or } y \in B\} = A \cup B.$$
    $$\cap X = \{y \mid \text{for all } Y \in X, y \in Y\} = \{y \mid y \in A \text{ and } y \in B\} = A \cap B.$$

    Another way to write this is $\cup\{A, B\} = A \cup B$ and $\cap\{A, B\} = A \cap B$.

2. Let A, B, and C be sets, and let $X = \{A, B, C\}$. Then

    $$\cup X = \{y \mid \text{there is } Y \in X \text{ with } y \in Y\} = \{y \mid y \in A, y \in B, \text{or } y \in C\} = A \cup B \cup C.$$
    $$\cap X = \{y \mid \text{for all } Y \in X, y \in Y\} = \{y \mid y \in A, y \in B, \text{and } y \in C\} = A \cap B \cap C.$$

    Another way to write this is $\cup\{A, B, C\} = A \cup B \cup C$ and $\cap\{A, B, C\} = A \cap B \cap C$.

3. Let $X = \{[0, r) \mid r \in \mathbb{R}^+\}$. Then we have $\cup X = [0, \infty)$ and $\cap X = \{0\}$.

    Another way to write this is $\cup\{[0, r) \mid r \in \mathbb{R}^+\} = [0, \infty)$ and $\cap\{[0, r) \mid r \in \mathbb{R}^+\} = \{0\}$.

Exercise 2.15: For each of the following, compute $\cup X$ and $\cap X$.

1. $X = \{\{a, b, c\}, \{b, c, d, e\}, \{a, b, x, y, z\}\}$ $\cup X =$ _____ $\cap X =$ _____

2. $X = \{(0, 4), [1, 5), [-17, 3), (2, \infty)\}$ $\cup X =$ _____ $\cap X =$ _____

3. $X = \{\{-n, \ldots, -3, -2, -1, 0, 1, 2, 3, 4, \ldots, n\} \mid n \in \mathbb{N}\}$ $\cup X =$ _____ $\cap X =$ _____

The Natural Numbers Revisited

Recall from Lesson 1 that the natural numbers are defined as follows:

$$0 = \emptyset$$
$$1 = \{\emptyset\} = \{0\}$$
$$2 = \{\emptyset, \{\emptyset\}\} = \{0, 1\}$$
$$3 = \{\emptyset, \{\emptyset\}, \{\emptyset, \{\emptyset\}\}\} = \{0, 1, 2\}$$
$$\ldots \qquad\qquad \ldots$$

In general, $n = \{0, 1, 2, \dots, n-1\}$ and $\mathbb{N} = \{0, 1, 2, \dots\}$

If n is a natural number, we define the **successor** of n, written n^+, to be the natural number $n^+ = n \cup \{n\}$.

Example 2.16:

1. $0^+ = 0 \cup \{0\} = \{0\} = 1$.
2. $1^+ = 1 \cup \{1\} = \{0\} \cup \{1\} = \{0, 1\} = 2$.
3. $2^+ = 2 \cup \{2\} = \{0, 1\} \cup \{2\} = \{0, 1, 2\} = 3$.

If n is a natural number such that $n \neq 0$, we define the **predecessor** of n, written n^-, to be the natural number k such that $n = k^+$. Thus, $n = n^- \cup \{n^-\}$.

Note that for all $n \in \mathbb{N}$, $(n^+)^- = n$ and for all $n \in \mathbb{N}$ with $n \neq 0$, $(n^-)^+ = n$.

Example 2.17:

1. Since $1 = 0^+$, $1^- = 0$.
2. Since $2 = 1^+$, $2^- = 1$.
3. Since $3 = 2^+$, $3^- = 2$.

Exercise 2.18: Find the natural number that each of the following expressions is equal to.

1. 3^+ _____
2. 3^- _____
3. $(3^+)^-$ _____
4. $(3^-)^+$ _____
5. $4 \cup \{4\}$ _____
6. $4 \cap \{4\}$ _____
7. $3 \cup 5$ _____
8. $3 \cap 5$ _____

Full solutions to these problems are available for free download here:

www.SATPrepGet800.com/STKZ3D

LEVEL 1

Let $A = \{x, y, z, w\}$ and $B = \{s, t, y, w\}$. Determine each of the following:

1. $A \cup B$

2. $A \cap B$

3. $A \setminus B$

4. $B \setminus A$

5. $A \triangle B$

Determine if each of the following sets is an interval.

6. $A = \{x \in \mathbb{R} \mid 12 \leq x \leq 15\}$

7. $B = \{x \in \mathbb{R} \mid x < -103\}$

8. $C = \{x \in \mathbb{Q} \mid x < -103\}$

9. $D = \mathbb{Q}^-$

10. $E = \mathbb{R}^+$

11. $F = \{x \in \mathbb{R} \mid x \geq -16\}$

12. $G = \{x \in \mathbb{R} \mid 0 \leq x < 999\}$

13. $H = \mathbb{R} \setminus \{0\}$

Sketch the graph of each of the following:

14. $\mathbb{R}$

15. $\mathbb{R}^+$

16. $\{-1, 1\}$

17. $(-1, 1)$

18. $[-1, \infty)$

19. $\mathbb{N}$

20. $\mathbb{Z}$

21. $(-\infty, -1)$

22. $(1, 2]$

LEVEL 2

Let $A = \left\{\emptyset, \{\emptyset, \{\emptyset\}\}\right\}$ and $B = \{\emptyset, \{\emptyset\}\}$. Compute each of the following:

23. $A \cup B$

24. $A \cap B$

25. $A \setminus B$

26. $B \setminus A$

27. $A \, \Delta \, B$

Let $A = (-17, 6)$ and $B = (-1, 17)$. Compute each of the following:

28. $A \cup B$

29. $A \cap B$

30. $A \setminus B$

31. $B \setminus A$

32. $A \, \Delta \, B$

Let $A = [14, \infty)$ and $B = (-\infty, 15)$. Compute each of the following:

33. $A \cup B$

34. $A \cap B$

35. $A \setminus B$

36. $B \setminus A$

37. $A \, \Delta \, B$

LEVEL 3

Let $A, B,$ and C be sets, let $X = (A \setminus B) \setminus C$, and let $Y = A \setminus (B \setminus C)$.

38. Draw Venn Diagrams for X and Y.

39. Is $X \subseteq Y$?

40. Is $Y \subseteq X$?

41. Is $X = Y$?

For each of the following, compute $\bigcup X$ and $\bigcap X$.

42. $X = \{\{0, 1, 2\}, \{1, 2, 3\}, \{2, 3, 4\}\}$

43. $X = \{\mathbb{N}, \mathbb{Z}, \mathbb{Q}, \mathbb{R}\}$

44. $X = \{(0, 10), (1, 11), (2, 12), (3, 13), (4, 14)\}$

45. $X = \{(-\infty, 20), [-5, 17), (4, 100]\}$

46. $X = \{(0, 1], (1, 2], (2, 3], (3, 4]\}$

Find the natural number that each of the following expressions is equal to.

47. 7^+

48. $12 \cup \{12\}$

49. $8 \cap 3$

37

LEVEL 4

For each of the following, compute $\cup X$ and $\cap X$.

50. $X = \{(0, q] \mid q \in \mathbb{Q}^+\}$

51. $X = \left\{ \left(\frac{1}{n}, 1 \right) \mid n \in \mathbb{Z}^+ \right\}$

52. $X = \{(n, n + 3) \mid n \in \mathbb{Z}\}$

53. $X = \left\{ \left(0, 1 + \frac{1}{n} \right) \mid n \in \mathbb{Z}^+ \right\}$

54. $X = \left\{ \left(-\infty, \frac{1}{n} \right] \mid n \in \mathbb{Z}^+ \right\}$

If X is a nonempty set of sets, we say that X is **disjoint** if $\cap X = \emptyset$. We say that X is **pairwise disjoint** if for all $A, B \in X$ with $A \neq B$, A and B are disjoint. For each of the following, determine if X is disjoint, pairwise disjoint, both, or neither.

55. $\{(n, n + 1) \mid n \in \mathbb{Z}\}$

56. $\{(n, n + 1] \mid n \in \mathbb{Z}\}$

57. $\left\{ \left(\frac{1}{n+1}, \frac{1}{n} \right) \mid n \in \mathbb{Z}^+ \right\}$

58. $\{\mathbb{Q}\}$

59. $\left\{ \mathbb{Q}^-, 2\mathbb{N}, \{n \in \mathbb{N} \mid n \text{ is a prime number greater than } 2\} \right\}$

LEVEL 5

Let A and B be sets with $B \subseteq A$. Determine if the following are true or false.

60. $A \cap B = A$

61. $A \setminus B \subseteq A$

Let X be a nonempty set of sets. Verify each of the following:

62. For all $A \in X$, $A \subseteq \bigcup X$.

63. For all $A \in X$, $\bigcap X \subseteq A$.

Let A be a set and let X be a nonempty set of sets. Verify each of the following:

64. $A \cap \bigcup X = \bigcup \{A \cap B \mid B \in X\}$.

65. $A \cup \bigcap X = \bigcap \{A \cup B \mid B \in X\}$.

66. $A \setminus \bigcup X = \bigcap \{A \setminus B \mid B \in X\}$.

67. $A \setminus \bigcap X = \bigcup \{A \setminus B \mid B \in X\}$.

Find the natural number that each of the following expressions is equal to or explain why the given expression is not a natural number.

68. $\bigcup \{2k \mid 0 \le k \le 100\}$

69. $\bigcup \{\{k\} \mid 0 \le k \le 100\}$

70. $\bigcup \{\{2k\} \mid 0 \le k \le 100\}$

71. $202 \setminus \bigcup \{2k \mid 0 \le k \le 100\}$

72. $(202 \setminus \bigcup \{2k \mid 0 \le k \le 100\}) \cup 201$

CHALLENGE PROBLEMS

73. Explain why $\mathcal{P}(A \cap B) = \mathcal{P}(A) \cap \mathcal{P}(B)$.

74. Determine conditions on sets A and B so that $\mathcal{P}(A \cup B) = \mathcal{P}(A) \cup \mathcal{P}(B)$.

LESSON 3
RELATIONS

Ordered Pairs

An **unordered pair** is a set with two elements. Recall, that a set doesn't change if we write the elements in a different order or if we write the same element multiple times. For example, $\{x, y\} = \{y, x\}$ and $\{x, x\} = \{x\}$.

There will be times when we would like for the order in which we write the elements to matter. In this case, instead of writing $\{x, y\}$, we will write (x, y). We call the expression (x, y) an **ordered pair**. It has the property that $(x, y) \neq (y, x)$. More specifically, we have that $(x, y) = (z, w)$ if and only if $x = z$ and $y = w$.

Note: Recall from Lesson 1 (see part 6 of Example 1.23) that the expression "if and only if" used above provides us with two pieces of information. It says "if $(x, y) = (z, w)$, then $x = z$ and $y = w$" and it also says "if $x = z$ and $y = w$, then $(x, y) = (z, w)$."

x and y are called the **coordinates** of the ordered pair (x, y).

Example 3.1:

1. $(0, 1)$ and $(1, 0)$ are different ordered pairs. That is, $(0, 1) \neq (1, 0)$. Both ordered pairs have the same coordinates, but these coordinates are in different positions. Compare this with the analogous unordered pairs $\{0, 1\}$ and $\{1, 0\}$, which are equal to each other.

2. $(0, 0)$ is an ordered pair with both coordinates equal to 0.

Exercise 3.2: Determine if each of the following is true or false.

1. $(a, b) = \{a, b\}$ _____

2. $(a, b) = (b, a)$ _____

3. $\{a, b\} = \{b, a\}$ _____

4. $\{a, a\} = (a, a)$ _____

Do ordered pairs actually exist? In other words, is there a set that has the defining property of an ordered pair? There are actually many ways to define such a set. One of the simplest ways to do this is as follows:

$$(x, y) = \{\{x\}, \{x, y\}\}$$

In Problems 70 – 76 below, you will be asked to show that with this definition, $(x, y) = (z, w)$ if and only if $x = z$ and $y = w$.

Example 3.3:

1. $(0, 1) = \{\{0\}, \{0, 1\}\}$. Observe that $(0, 1)$ is a set with two elements, namely $\{0\}$ and $\{0, 1\}$. Note that 0 and 1 are **not** elements of $(0, 1)$.

2. $(0, 0) = \{\{0\}, \{0, 0\}\} = \{\{0\}, \{0\}\} = \{\{0\}\}$. Observe that $(0, 0)$ is a set with just one element, namely $\{0\}$. Note that 0 is **not** an element of $(0, 0)$.

Exercise 3.4: Determine x and y so that each of the following equations is true, or state that the equation has no such solution.

1. $(x, y) = (2, 7)$ _____

2. $\{x, y\} = \{3\}$ _____

3. $(x, y) = \{\{1\}, \{1, 5\}\}$ _____

4. $(x, y) = \{\{1, 5\}, \{5\}\}$ _____

5. $(x, y) = \{\{0\}, \{1\}\}$ _____

6. $(x, y) = \{4\}$ _____

7. $(x, y) = \{\{9\}\}$ _____

8. $(x, y) = \{5, \{5, 6\}\}$ _____

Note: (x, y) is formally treated as an abbreviation for the set $\{\{x\}, \{x, y\}\}$. In the study of set theory, every object can be written as a set like this. It's often convenient to use abbreviations, but we should always be aware that if necessary, we can write any object in its unabbreviated form.

We can extend the idea of an ordered pair to an **ordered k-tuple**. An ordered 3-tuple (also called an **ordered triple**) is defined by $(x, y, z) = ((x, y), z)$, an ordered 4-tuple is $(x, y, z, w) = ((x, y, z), w)$, and so on. For a general k-tuple, we will use a single letter with **subscripts** for the variable names. For example, using the letter x, we can write a 5-tuple as $(x_1, x_2, x_3, x_4, x_5)$. More generally, we can write a k-tuple as $(x_1, x_2, \ldots, x_k)$.

In Problems 77 and 78 below, you will be asked to verify that $(x, y, z) = (u, v, w)$ if and only if $x = u$, $y = v$, and $z = w$. This result extends in the obvious way to k-tuples in general.

Exercise 3.5: Write each of the following in its unabbreviated form.

1. (x, y, z) _____

2. (x, x, x) _____

Cartesian Products

The **Cartesian product** of the sets A and B, written $A \times B$ is the set of ordered pairs (a, b) with $a \in A$ and $b \in B$. Symbolically, we have

$$A \times B = \{(a, b) \mid a \in A \wedge b \in B\}.$$

Observe that if A and B are finite sets with $|A| = m$ and $|B| = n$, then $|A \times B| = mn$.

Note: Recall from Note 4 following Example 1.13 that the symbol "∧" is called a **conjunction symbol** and it can be read as "and." Also, recall from lesson 1 that $|A|$ is the **cardinality** of the set A.

Example 3.6:

1. Let $A = \{0, 1\}$ and $B = \{2, 3, 4\}$. Then $A \times B = \{(0, 2), (0, 3), (0, 4), (1, 2), (1, 3), (1, 4)\}$. Note that $|A| = 2$, $|B| = 3$, and $|A \times B| = 2 \cdot 3 = 6$.

2. Let $C = \emptyset$ and $D = \{a, b, c, d\}$. Then $C \times D = \emptyset$ (since there are no elements in C, there can be no elements in $C \times D$). Note that $|C| = 0$, $|D| = 4$, and $|C \times D| = 0 \cdot 4 = 0$.

3. $\mathbb{N} \times \mathbb{Z} = \{(m, n) \mid m \in \mathbb{N} \wedge n \in \mathbb{Z}\}$. For example, $(5, -3) \in \mathbb{N} \times \mathbb{Z}$, whereas $(-3, 5) \notin \mathbb{N} \times \mathbb{Z}$ (although it is in $\mathbb{Z} \times \mathbb{N}$). We can visualize $\mathbb{N} \times \mathbb{Z}$ as follows:

$$\ldots, (0, -3), (0, -2), (0, -1), (0, 0), (0, 1), (0, 2), (0, 3), \ldots$$
$$\ldots, (1, -3), (1, -2), (1, -1), (1, 0), (1, 1), (1, 2), (1, 3), \ldots$$
$$\ldots, (2, -3), (2, -2), (2, -1), (2, 0), (2, 1), (2, 2), (2, 3), \ldots$$
$$\vdots \qquad\qquad \vdots \qquad\qquad \vdots$$

4. $\mathbb{R} \times \mathbb{R} = \{(x, y) \mid x, y \in \mathbb{R}\}$. We can visualize elements of $\mathbb{R} \times \mathbb{R}$ as points in the **Cartesian plane.** A portion of the Cartesian plane is shown to the right. The elements $(3, 2)$ and $(-1, -2)$ of $\mathbb{R} \times \mathbb{R}$ are displayed as points.

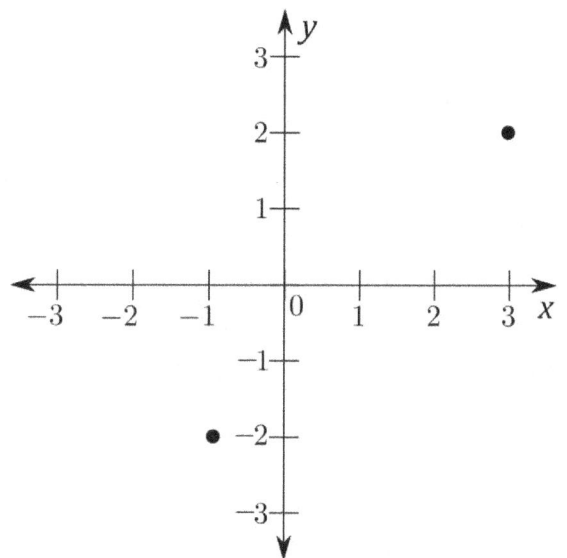

    We form the Cartesian plane by taking two copies of the real line and placing one horizontally and the other vertically, exactly as we did for the Complex Plane in Lesson 1. The horizontal copy of the real line is called the x-axis (labeled x in the figure) and the vertical copy of the real line is called the y-axis (labeled y in the figure). The two axes intersect at the point $(0, 0)$. This point is called the **origin**.

    Notice that visually the Cartesian plane $\mathbb{R} \times \mathbb{R}$ is indistinguishable from the Complex Plane.

Exercise 3.7: Determine if each of the following is true or false.

1. $(0, 0) \in \mathbb{N} \times \mathbb{N}$ _____

2. $\{0, 0\} \in \mathbb{N} \times \mathbb{N}$ _____

3. $\{\{0\}\} \in \mathbb{N} \times \mathbb{N}$ _____

4. $\left(1 + \frac{2}{3}i, 2.7\right) \in \mathbb{C} \times \mathbb{Q}$ _____

5. $\left(1 + \frac{2}{3}i, 2.7\right) \in \mathbb{Q} \times \mathbb{C}$ _____

6. $|\{a, b, c, d\} \times \{\alpha, \beta, \gamma\}| = 7$ _____

We can extend the definition of the Cartesian product to more than two sets in the obvious way:

$$A \times B \times C = \{(a, b, c) \mid a \in A \land b \in B \land c \in C\}$$

$$A \times B \times C \times D = \{(a, b, c, d) \mid a \in A \land b \in B \land c \in C \land d \in D\}$$

Observe that if A, B, and C are finite sets with $|A| = m$, $|B| = n$, and $|C| = k$, then we have $|A \times B \times C| = mnk$.

Similarly, if A, B, C, and D are finite sets with $|A| = m$, $|B| = n$, $|C| = k$, and $|D| = t$, then we have $|A \times B \times C \times D| = mnkt$.

In general, the **Cartesian product** of the sets $A_1, A_2, \ldots, A_n$ is

$$A_1 \times A_2 \times \cdots \times A_n = \{(a_1, a_2, \ldots, a_n) \mid a_1 \in A_1 \land a_2 \in A_2 \land \cdots \land a_n \in A_n\}.$$

If $A_1, A_2, \ldots, A_n$ are finite, then the cardinality of $A_1 \times A_2 \times \cdots \times A_n$ is the product of the cardinalities of $A_1, A_2, \ldots, A_n$. Symbolically, we have $|A_1 \times A_2 \times \cdots \times A_n| = |A_1| \cdot |A_2| \cdots |A_n|$.

Example 3.8:

1. $\{x\} \times \{y\} \times \{z\} \times \{w\} = \{(x, y, z, w)\}$.

   Note that $|\{x\} \times \{y\} \times \{z\} \times \{w\}| = |\{x\}| \cdot |\{y\}| \cdot |\{z\}| \cdot |\{w\}| = 1 \cdot 1 \cdot 1 \cdot 1 = 1$.

2. $\{0\} \times \{0,1\} \times \{1\} \times \{0,1\} \times \{0\} = \{(0,0,1,0,0), (0,0,1,1,0), (0,1,1,0,0), (0,1,1,1,0)\}$.

   Note that $|\{0\} \times \{0,1\} \times \{1\} \times \{0,1\} \times \{0\}| = 1 \cdot 2 \cdot 1 \cdot 2 \cdot 1 = 4$.

3. $\{0\} \times \mathbb{Z} \times \mathbb{N} \times \mathbb{R} = \{(0, m, n, x) \mid m \in \mathbb{Z} \land n \in \mathbb{N} \land x \in \mathbb{R}\}$.

Exercise 3.9: Let $A = \{0\}$, $B = \{1, 2\}$, $C = \{0, 1, 2\}$. And $D = \{2\}$. Use the roster method to describe the set $A \times B \times C \times D$.

We abbreviate Cartesian products of sets with themselves using exponents.

$$A^2 = A \times A \qquad A^3 = A \times A \times A \qquad A^4 = A \times A \times A \times A \qquad A^n = \underbrace{A \times A \times \cdots \times A}_{n \text{ times}}$$

Example 3.10:

1. $\mathbb{Z}^2 = \mathbb{Z} \times \mathbb{Z} = \{(x, y) \mid x, y \in \mathbb{Z}\}$ is the set of ordered pairs of integers. A few sample elements in $\mathbb{Z}^2$ are $(0, 0)$, $(-1, 2)$, $(15, -106)$ and $(-53, -53)$.

2. $\mathbb{N}^5 = \mathbb{N} \times \mathbb{N} \times \mathbb{N} \times \mathbb{N} \times \mathbb{N} = \{(a, b, c, d, e) \mid a, b, c, d, e \in \mathbb{N}\}$ is the set of ordered 5-tuples of natural numbers. A few sample elements in $\mathbb{N}^5$ are $(0,0,0,0,0)$, $(1,1,2,3,3)$, $(0,1,17,86,0)$ and $(1000, 2529, 8, 900, 106)$.

3. $\{0,1\}^2 = \{0,1\} \times \{0,1\} = \{(0,0), (0,1), (1,0), (1,1)\}$.

4. $\{0,1\}^3 = \{0,1\} \times \{0,1\} \times \{0,1\}$

   $$= \{(0,0,0), (0,0,1), (0,1,0), (0,1,1), (1,0,0), (1,0,1), (1,1,0), (1,1,1)\}.$$

Exercise 3.11: Use set-builder notation to describe each of the following sets.

1. $\mathbb{R}^2$
2. $\mathbb{Z}^3$
3. $\mathbb{C}^5$
4. $\mathbb{R}^n$ (where $n \in \mathbb{Z}^+$)
5. $\mathbb{C}^n$ (where $n \in \mathbb{Z}^+$)

Binary Relations

A **relation** on a set describes a relationship among the set's objects. In this section, we will be interested in **binary relations** (the prefix "bi" means "two"). The idea is that given **two** objects from a set, either these two objects satisfy the given relationship or they do not.

Example 3.12:

1. "=" (equals) is a binary relation on the set of natural numbers. The statement $0 = 0$ (pronounced "0 is equal to 0") is true, whereas the statement $0 = 1$ is false. Notice that for natural numbers a and b, the statement $a = b$ is true if and only if a and b are the same natural number. Instead of saying "$a = b$ is true," we will simply write $a = b$. Instead of saying "$a = b$ is false," we will write $a \neq b$.

   More generally, if A is any set, then "=" is a binary relation on A. Once again, $a = b$ if and only if a and b are the same element of A.

2. Let $A = \{a, b, c\}$. Recall that the power set of A is

   $$\mathcal{P}(A) = \{\emptyset, \{a\}, \{b\}, \{c\}, \{a, b\}, \{a, c\}, \{b, c\}, \{a, b, c\}\}.$$

   "$\subseteq$" (subset) is a binary relation on $\mathcal{P}(A)$. We have for example $\{a\} \subseteq \{a, c\}$ (pronounced "$\{a\}$ is a subset of $\{a, c\}$"), whereas $\{a\} \nsubseteq \{b, c\}$.

   More generally, if X is any set, then "$\subseteq$" is a binary relation on $\mathcal{P}(X)$.

3. "$\leq$" (less than or equal to) is a binary relation on the set of natural numbers. We have for example $2 \leq 7$ (pronounced "2 is less than or equal to 7") and $2 \leq 2$, whereas $2 \nleq 1$.

   "$\leq$" is also a binary relation on many other sets such as the integers, the rational numbers, and the real numbers.

Exercise 3.13: Consider the binary relations "$<$" (less than) and "$>$" (greater than) on the set of integers. Determine if each of the following statements is true or false.

1. $0 < -1$ _____
2. $-1 > -3$ _____
3. $-1 \nless -5$ _____
4. $5 \ngtr -5$ _____

Let's now give a more formal definition of a binary relation.

A **binary relation** on a set A is a subset of $A^2 = A \times A$. Symbolically, we have

$$R \text{ is a binary relation on } A \text{ if and only if } R \subseteq A \times A.$$

We will usually abbreviate $(a, b) \in R$ as aRb, as we have done repeatedly in Example 3.12 above.

For example, formally speaking, the relation "=" on $\mathbb{N}$ is the set $E = \{(a, a) \mid a \in \mathbb{N}\}$. Observe that $E \subseteq \mathbb{N} \times \mathbb{N}$. We have $(0, 0) \in E$, whereas $(0, 1) \notin E$. We abbreviate the expression $(0, 0) \in E$ by $0E0$, or better yet, $0 = 0$. Similarly, we abbreviate the expression $(0, 1) \notin E$ by $0 \neq 1$.

Notes: (1) We should have really used the symbol "=" instead of the letter "E" above. However, the expression $== \{(a, a) \mid a \in \mathbb{N}\}$ looks particularly confusing with the two equal signs placed next to each other. In fact, it could easily be mistaken for a typo. This is the only reason I decided to change the symbol "=" to the letter "E" when writing the relation in its unabbreviated form.

(2) The statement $R \subseteq A \times A$ is equivalent to the statement $R \in \mathcal{P}(A \times A)$ (see Lesson 1). It follows that for a finite set A, the number of binary relations on A is $|\mathcal{P}(A \times A)|$.

Example 3.14:

1. Let $R = \{(a, b) \in \mathbb{N} \times \mathbb{N} \mid a < b\}$. For example, we have $(0, 1) \in R$ because $0 < 1$. However, $(1, 1) \notin R$ because $1 \not< 1$. We abbreviate $(0, 1) \in R$ by $0R1$.

    Observe that $R \subseteq \mathbb{N} \times \mathbb{N}$, and so, R is a binary relation on $\mathbb{N}$.

    We would normally use the name $<$ for this relation R. So, we have $(0, 1) \in <$, which we abbreviate as $0 < 1$, and we have $(1, 1) \notin <$, which we abbreviate as $1 \not< 1$.

2. There are binary relations $<, \leq, >, \geq$ defined on $\mathbb{N}, \mathbb{Z}, \mathbb{Q}$, and $\mathbb{R}$. For example, if we consider $> \subseteq \mathbb{Z}^2$, we have $(13, -7) \in >$, or equivalently, $13 > -7$.

3. Let $A = \{a\}$. Since $|A| = 1$, we have $|A \times A| = 1 \cdot 1 = 1$. So, $|\mathcal{P}(A \times A)| = 2^1 = 2$. So, there are 2 binary relations on A. They are $R_1 = \emptyset$ and $R_2 = \{(a, a)\}$.

4. Let A be a set and let R be the binary relation on A defined by $R = \{(a, b) \in A \times A \mid a \in b\}$. R is known as the **membership relation**, and it is usually denoted by $\in$. So, if $a, b \in A$ and a is a member of b, we can write $(a, b) \in \in$, which we will usually abbreviate as $a \in b$. As a specific example, let $A = \{\emptyset, \{\emptyset\}, \{\{\emptyset\}\}\}$. Then $(\emptyset, \{\emptyset\}) \in \in$, or equivalently, $\emptyset \in \{\emptyset\}$. Similarly, we have $(\{\emptyset\}, \{\{\emptyset\}\}) \in \in$, or equivalently, $\{\emptyset\} \in \{\{\emptyset\}\}$.

5. Let $R = \{((a, b), (c, d)) \in (\mathbb{N} \times \mathbb{N})^2 \mid a + d = b + c\}$. Then R is a binary relation on $\mathbb{N} \times \mathbb{N}$. For example, we have $(5, 0)R(6, 1)$ because $5 + 1 = 0 + 6$. However, we see that $(5, 0)\not\!R(6, 2)$ because $5 + 2 \neq 0 + 6$.

6. Let $R = \{((a, b), (c, d)) \in (\mathbb{Z} \times \mathbb{Z}^*)^2 \mid ad = bc\}$. (Recall that $\mathbb{Z}^*$ is the set of *nonzero* integers.) Then R is a binary relation on $\mathbb{Z} \times \mathbb{Z}^*$. For example, $(1, 2)R(2, 4)$ because $1 \cdot 4 = 2 \cdot 2$. However, $(1, 2)\not\!R(2, 5)$ because $1 \cdot 5 \neq 2 \cdot 2$. Compare this to the rational number system (see Lesson 1), where we have $\frac{1}{2} = \frac{2}{4}$ because $1 \cdot 4 = 2 \cdot 2$, but $\frac{1}{2} \neq \frac{2}{5}$ because $1 \cdot 5 \neq 2 \cdot 2$.

Exercise 3.15: Let $X = \{0, 1\}$.

1. Compute $|X \times X|$ _____

2. How many binary relations are there on X?   _____

3. List all binary relations on X.

_____

_____

_____

The **domain** of a binary relation R is the set of all possible "inputs" of the relation. In other words, if $(x, y) \in R$ (or equivalently, xRy) for some y, then x is in the domain of R. Formally, we write

$$\text{dom } R = \{x \mid \exists y(xRy)\}.$$

The symbol $\exists$ is called an **existential quantifier**, and it is pronounced "There exists" or "There is." So, the expression $\exists y(xRy)$ can be translated into English as "There exists a y such that xRy."

Similarly, the **range** of a binary relation R, written $\text{ran } R$, is the set of all possible "outputs" of the relation. In other words, if $(x, y) \in R$ (or equivalently, xRy) for some x, then y is in the range of R. Formally, we write

$$\text{ran } R = \{y \mid \exists x(xRy)\}.$$

The expression $\exists x(xRy)$ can be translated into English as "There exists an x such that xRy."

The **field** of a binary relation R is $\text{dom } R \cup \text{ran } R$. In other words, the field of R consists of all elements that are in the domain of R, the range of R, or both.

Example 3.16:

1. Let $R = \{(a, b) \in \mathbb{N} \times \mathbb{N} \mid a < b\}$. Then $\text{dom } R = \mathbb{N}$, $\text{ran } R = \mathbb{N}$, and field $R = \mathbb{N} \cup \mathbb{N} = \mathbb{N}$.

2. Let $B = \{0, 1, 2, 3\}$ and $R = \{(0, 2), (0, 3), (1, 3)\}$. Then $\text{dom } R = \{0, 1\}$, $\text{ran } R = \{2, 3\}$, and field $R = \{0, 1\} \cup \{2, 3\} = \{0, 1, 2, 3\} = B$.

3. Let $R = \{((a, b), (c, d)) \in (\mathbb{N} \times \mathbb{N})^2 \mid a + d = b + c\}$. Then $\text{dom } R = \mathbb{N} \times \mathbb{N}$, $\text{ran } R = \mathbb{N} \times \mathbb{N}$, and field $R = (\mathbb{N} \times \mathbb{N}) \cup (\mathbb{N} \times \mathbb{N}) = \mathbb{N} \times \mathbb{N}$.

Exercise 3.17: Find the domain, range, and field of each of the following binary relations:

1. $R = \{(0, c), (1, a), (5, b), (9, d)\}$

2. $S = \mathbb{Z} \times \mathbb{C}$

3. $T = \{((a, b), (c, d)) \in (\mathbb{Z} \times \mathbb{Z}^*)^2 \mid ad = bc\}$

We say that a binary relation R on a set A is

- **reflexive** if for all $a \in A$, aRa.

- **symmetric** if for all $a, b \in A$, aRb implies bRa.

- **transitive** if for all $a, b, c \in A$, aRb and bRc imply aRc.

- **antireflexive** if for all $a \in A$, $a\not\!Ra$.

- **antisymmetric** if for all $a, b \in A$, aRb and bRa imply $a = b$.

- **trichotomous** if for all $a, b \in A$, exactly one of aRb, bRa, or $a = b$ holds.

Example 3.18:

1. Let A be a nonempty set and consider the binary relation "=" on A. This relation is reflexive ($a = a$), symmetric (if $a = b$, then $b = a$), transitive (if $a = b$ and $b = c$, then $a = c$), and antisymmetric (trivially). This relation is **not** antireflexive because $a \neq a$ is false for any $a \in A$. This relation is **not** trichotomous because if $a \neq b$, then $b \neq a$.

2. Let A be a set with at least two elements and consider the binary relation "$\subseteq$" on $\mathcal{P}(A)$. This relation is reflexive (every set is a subset of itself), transitive (if $X \subseteq Y$ and $Y \subseteq Z$, then $X \subseteq Z$), and antisymmetric (if $X \subseteq Y$ and $Y \subseteq X$, then $X = Y$). This relation is **not** symmetric (for example, if a and b are distinct, then $\{a\} \not\subseteq \{b\}$ and $\{b\} \not\subseteq \{a\}$), it is **not** antireflexive ($X \not\subseteq X$ is false for any set X), and it is **not** trichotomous (for example, if a and b are distinct, then $\{a\} \not\subseteq \{b\}$, $\{b\} \not\subseteq \{a\}$, and $\{a\} \neq \{b\}$).

3. Consider the binary relation "$\leq$" on $\mathbb{N}$. This relation is reflexive ($a \leq a$), transitive (if $a \leq b$ and $b \leq c$, then $a \leq c$), and antisymmetric (if $a \leq b$ and $b \leq a$, then $a = b$). This relation is **not** symmetric (for example, $0 \leq 1$, but $1 \not\leq 0$), it is **not** antireflexive (for example $0 \not\leq 0$ is false), and it is **not** trichotomous ($0 \leq 0$ and $0 = 0$ both hold).

4. Let $R = \{(0,0), (0,2), (2,0), (2,2), (2,3), (3,2), (3,3)\}$ be a binary relation on $\mathbb{N}$. R is **not** reflexive because $1 \in \mathbb{N}$, but $(1,1) \notin R$ (however, if we were to consider R as a relation on $\{0, 2, 3\}$ instead of on $\mathbb{N}$, then R **would** be reflexive). We can see that R is symmetric by observation. R is **not** transitive because we have $(0,2), (2,3) \in R$, but $(0,3) \notin R$. R is **not** antisymmetric because we have $(2,3), (3,2) \in R$ and $2 \neq 3$. R is **not** antireflexive because $(0,0) \in R$ (and also, $(2,2) \in R$ and $(3,3) \in R$). Finally, R is not trichotomous because $(2,3)$ and $(3,2)$ are both in R (and for several other reasons as well).

Exercise 3.19: Consider the relation $>$ on $\mathbb{Z}$.

1. Is $>$ reflexive on $\mathbb{Z}$? _____

2. Is $>$ symmetric on $\mathbb{Z}$? _____

3. Is $>$ transitive on $\mathbb{Z}$? _____

4. Is $>$ antireflexive on $\mathbb{Z}$? _____

5. Is $>$ antisymmetric on $\mathbb{Z}$? _____

6. Is $>$ trichotomous on $\mathbb{Z}$? _____

n-ary Relations

We can extend the idea of a binary relation on a set A to an **n-ary relation** on A. For example, a 3-ary relation (or **ternary relation**) on A is a subset of $A^3 = A \times A \times A$. More generally, we have that R is an n-ary relation on A if and only if $R \subseteq A^n$. A **1-ary relation** (or **unary relation**) on A is just a subset of A.

Example 3.20:

1. $\mathbb{R}$ is a unary relation on $\mathbb{C}$ because $\mathbb{R} \subseteq \mathbb{C}$.

2. Let $R = \{(x, y, z) \in \mathbb{Z}^3 \mid x + y = z\}$. Then R is a ternary (or 3-ary) relation on $\mathbb{Z}$. We have, for example, $(1, 2, 3) \in R$ (because $1 + 2 = 3$) and $(1, 2, 4) \notin R$ (because $1 + 2 \neq 4$).

3. Let C be the set of all colors. For example, blue $\in C$, pink $\in C$, and violet $\in C$. Let $S = \{(a, b, c) \in C^3 \mid$ when a and b are combined in equal quantities, the result is $c\}$. Then S is a ternary relation on C. We have, for example, (red, yellow, orange) $\in S$.

4. Let $T = \{(a, b, c, d, e) \in \mathbb{N}^5 \mid ab + c = de\}$. Then T is a 5-ary relation on $\mathbb{N}$. We have, for example, $(1, 2, 8, 5, 2) \in T$ ($1 \cdot 2 + 8 = 5 \cdot 2$) and $(1, 1, 1, 1, 1) \notin T$ ($1 \cdot 1 + 1 \neq 1 \cdot 1$).

Exercise 3.21: Let $X = \{0, 1\}$.

1. Compute $|X^3|$ _____

2. How many ternary relations are there on X? _____

3. How many 4-ary relations are there on X? _____

4. Let $R = \{(x, y, z, w) \in X^4 \mid xyzw = 1\}$. List the elements of R.

Ordered Sets

A binary relation $<$ on a set A is a **strict linear ordering** on A if $<$ is transitive and trichotomous on A. In this case, we will call $(A, <)$ an **ordered set**.

Example 3.22:

1. The binary relation $<$ on $\mathbb{N}$ is a strict linear ordering. Therefore, $(\mathbb{N}, <)$ is an ordered set.

2. The binary relation $<$ on $\mathbb{Z}$ is a strict linear ordering. Therefore, $(\mathbb{Z}, <)$ is an ordered set.

3. The binary relation $<$ on $\mathbb{Q}$ is a strict linear ordering. Here we have $\frac{a}{b} < \frac{c}{d}$ if and only if $ad < bc$. Therefore, $(\mathbb{Q}, <)$ is an ordered set.

4. The binary relation $=$ is **not** a strict linear ordering on any nonempty set A because it is not trichotomous on A (see part 1 of Example 3.18). Therefore, $(A, =)$ is **not** an ordered set.

5. Let X be the set of planets in our solar system. Define a binary relation $\lhd$ on X by $x \lhd y$ if x is closer to the sun than y. Then $\lhd$ is a strict linear ordering on X. Therefore, $(X, \lhd)$ is an ordered set.

Exercise 3.23: Determine if each of the following is an ordered set.

1. $(\mathbb{N}, \leq)$ _____

2. $(\mathbb{Z}, \leq)$ _____

3. $(\mathbb{N}, >)$ _____

4. $(\mathbb{Z}, >)$ _____

5. $(\mathcal{P}(A), \subset)$, where A is a set with at least two elements and $\subset$ is the proper subset relation.

_____

If $(A, <)$ is an ordered set (so that $<$ is a strict linear ordering on A), then we define the binary relation $\leq$ on A by $a \leq b$ if and only if $a < b$ or $a = b$. In this case, $\leq$ is called a **linear ordering** on A.

Exercise 3.24: Let $(A, <)$ be an ordered set (so that $<$ is a strict linear ordering on A). Verify each of the following statements about the corresponding linear ordering $\leq$.

1. $\leq$ is **not** trichotomous on A. _____

2. $\leq$ is reflexive on A. _____

3. $\leq$ is antisymmetric on A. _____

4. $\leq$ is transitive on A. _____

5. $\leq$ satisfies the following **comparability condition** on A: if $a, b \in A$, then $a \leq b$ or $b \leq a$.

_____

If $(A, <)$ is an ordered set, we may write $b > a$ in place of $a < b$. We may also write $b \geq a$ in place of $a \leq b$.

A binary relation $<$ on a set A is a **strict partial ordering** on A if $<$ is antireflexive, antisymmetric, and transitive on A. In this case, we will call $(A, <)$ a **strict partially ordered set** (which we often abbreviate as **strict poset**).

Example 3.25:

1. Every strict linear ordering $<$ on a set A is a strict partial ordering on A. Indeed, since $<$ is trichotomous on A, we cannot have $a < a$ (so that $<$ is antireflexive on A) and we cannot have both $a < b$ and $b < a$ (so that $<$ is antisymmetric on A). It follows that every ordered set is a strict poset.

2. If A is a set with at least two elements, then $(\mathcal{P}(A), \subset)$ is a strict poset that is **not** an ordered set (the relation here is the **proper subset** relation). The relation $\subset$ is antireflexive, antisymmetric, and transitive, but it is not trichotomous. Indeed, if $a, b \in A$ with $a \neq b$, then we have $\{a\} \not\subset \{b\}$ and $\{b\} \not\subset \{a\}$.

3. Let $X = \{x \mid x \text{ is a word in the English language}\}$ and define the **dictionary order** on X as follows: $x <_D y$ if x appears before y alphabetically. Then $(X, <_D)$ is a strict poset. In this strict poset, we have aardvark $<_D$ antelope (because a = a and a appears before n alphabetically), we have stranger $<_D$ violin (because s appears before v alphabetically), and we have dragon $<_D$ drainage (because d = d, r = r, a = a and g appears before i alphabetically).

We can use a similar idea to define the **dictionary order** on products of strict posets.

For example, the dictionary order $<_D$ can be defined on $\mathbb{Z} \times \mathbb{Q}$ by $(a, b) <_D (c, d)$ if and only if either $a <_{\mathbb{Z}} c$ or both $a = c$ and $b <_{\mathbb{Q}} d$, where $<_{\mathbb{Z}}$ and $<_{\mathbb{Q}}$ are the usual strict partial orderings on $\mathbb{Z}$ and $\mathbb{Q}$, respectively. It is straightforward (but a bit tedious) to verify that $(\mathbb{Z} \times \mathbb{Q}, <_D)$ is a strict poset. In fact, $(\mathbb{Z} \times \mathbb{Q}, <_D)$ is an ordered set because $(\mathbb{Z}, <_{\mathbb{Z}})$ and $(\mathbb{Q}, <_{\mathbb{Q}})$ are both ordered sets.

In the ordered set $(\mathbb{Z} \times \mathbb{Q}, <_D)$, we have $(4, 7) <_D (9, -2)$ because $4 <_{\mathbb{Z}} 9$ (notice that the second coordinates are irrelevant because the first coordinates are not equal). We also have $\left(3, \frac{1}{2}\right) <_D \left(3, \frac{3}{4}\right)$ because $3 = 3$ and $\frac{1}{2} <_{\mathbb{Q}} \frac{3}{4}$.

We can visualize this particular dictionary order in a Cartesian plane as solid vertical lines passing through each integer value along the x-axis. Each point is less than any point higher than it on the same vertical line and each point is also less than any point on a vertical line to the right of that point (regardless of the height).

In the figure below, we see that $(1, -2) <_D \left(1, \frac{3}{2}\right)$ because $\left(1, \frac{3}{2}\right)$ is above $(1, -2)$ on the same vertical line. We also have $(1, -2) <_D (3, -3)$ because $(3, -3)$ is to the right of $(1, -2)$ (note that it does **not** matter that $(3, -3)$ is below $(1, -2)$).

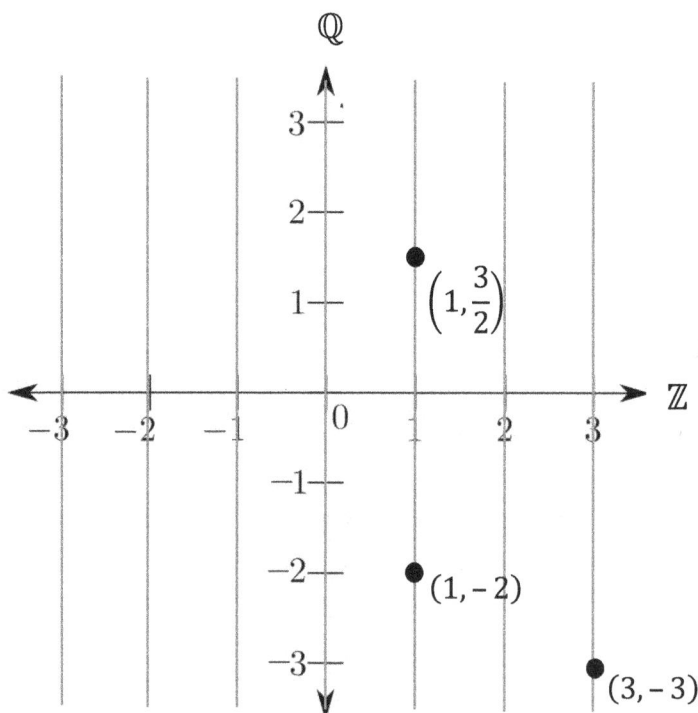

As another example, in the dictionary order of $\mathbb{R}^5$, we have $(1, 3, 5, 2, 11) <_D (1, 3, 5, 4, 10)$ because $1 = 1, 3 = 3, 5 = 5$, and $2 <_{\mathbb{R}} 4$.

50

As one more example, let $A = \{a, b\}$, consider the strict poset $(\mathcal{P}(A), \subset)$ (the relation here is the **proper subset** relation), and let $<_D$ be the corresponding dictionary order on $(\mathcal{P}(A))^2 = \mathcal{P}(A) \times \mathcal{P}(A)$. We have $(\emptyset, \{a\}) <_D (\{b\}, \{b\})$ because $\emptyset \subset \{b\}$. Also, we have $(\{a\}, \{b\}) <_D (\{a\}, \{a, b\})$ because $\{a\} = \{a\}$ and $\{b\} \subset \{a, b\}$. Note that in this case, we have a strict poset that is **not** an ordered set. Indeed, trichotomy is violated by $(\{a\}, \emptyset)$ and $(\{b\}, \emptyset)$ because $\{a\} \not\subset \{b\}$ and $\{b\} \not\subset \{a\}$, just like we observed in part 2 of this example above. As another example, trichotomy is also violated by $(\{a\}, \{a\})$ and $(\{a\}, \{b\})$.

Exercise 3.26: Determine if each of the following is a strict poset.

1. $(\mathbb{N}, >)$ _____

2. $(\mathbb{Z}, \geq)$ _____

3. $(\mathbb{Q}, <)$ _____

4. $(\{x, y, z\}, \{(x, y), (x, z), (y, z)\})$, where $x, y,$ and z are distinct

_____

If $(A, <)$ is a strict partially ordered set (so that $<$ is a strict partial ordering on A), then we define the binary relation $\leq$ on A by $a \leq b$ if and only if $a < b$ or $a = b$. In this case, $\leq$ is called a **partial ordering** on A and we will call $(A, \leq)$ a **partially ordered set** or **poset**.

Exercise 3.27: Let $(A, <)$ be a strict partially ordered set (so that $<$ is a strict partial ordering on A). Verify each of the following statements about the corresponding partial ordering $\leq$.

1. $\leq$ is reflexive on A. _____

2. $\leq$ is antisymmetric on A. _____

3. $\leq$ is transitive on A. _____

The Natural Numbers as an Ordered Set

Recall the more formal definition of the natural numbers:

$$0 = \emptyset$$
$$1 = \{\emptyset\} = \{0\}$$
$$2 = \{\emptyset, \{\emptyset\}\} = \{0, 1\}$$
$$3 = \{\emptyset, \{\emptyset\}, \{\emptyset, \{\emptyset\}\}\} = \{0, 1, 2\}$$
$$\cdots \qquad \cdots$$

In general, we let $n = \{0, 1, 2, \ldots, n - 1\}$ and we define the **natural numbers** to be the set

$$\mathbb{N} = \{0, 1, 2, 3, 4, \ldots\}$$

If n is a natural number, the **successor** of n is $n^+ = n \cup \{n\}$. Note that $n^+ = \{0, 1, 2, \ldots, n - 1, n\}$.

If n is a natural number such that $n \neq 0$, the **predecessor** of n, written n^-, is the natural number k such that $n = k^+$.

We define a binary relation $<_\mathbb{N}$ on $\mathbb{N}$ by $n <_\mathbb{N} m$ if and only if $n \in m$. We will usually abbreviate $<_\mathbb{N}$ simply by $<$, especially if it is already clear that we are working with the natural numbers.

With this definition, $(\mathbb{N}, <_\mathbb{N})$ is an ordered set. Below are a few different ways we can visualize the ordering $<_\mathbb{N} = \in$.

$$0 \in 1 \in 2 \in 3 \in 4 \in \cdots \in n \in \cdots$$

$$0 \in \{0\} \in \{0, 1\} \in \{0, 1, 2\} \in \{0, 1, 2, 3\} \in \cdots \in \{0, 1, 2, \ldots, n-1\} \in \cdots$$

$$\emptyset \in \{\emptyset\} \in \{\emptyset, \{\emptyset\}\} \in \big\{\emptyset, \{\emptyset\}, \{\emptyset, \{\emptyset\}\}\big\} \in \Big\{\emptyset, \{\emptyset\}, \{\emptyset, \{\emptyset\}\}, \big\{\emptyset, \{\emptyset\}, \{\emptyset, \{\emptyset\}\}\big\}\Big\} \in \cdots$$

$$0 <_\mathbb{N} 1 <_\mathbb{N} 2 <_\mathbb{N} 3 <_\mathbb{N} 4 <_\mathbb{N} \cdots <_\mathbb{N} n <_\mathbb{N} \cdots$$

$$0 < 1 < 2 < 3 < 4 < \cdots < n < \cdots$$

Note that each of the five visualizations above represent different ways of expressing the same thing.

Example 3.28:

1. $0 \in \{0\}$. Therefore, $0 \in 0 \cup \{0\} = 1$. So, $0 < 1$.

2. $1 = \{0\} \notin \emptyset = 0$. So, $1 \not< 0$.

3. $1 = \{0\} \notin 0$ and $1 = \{0\} \notin \{0\}$. Therefore, $1 = \{0\} \notin 0 \cup \{0\} = 1$. So, $1 \not< 1$.

4. If $n \neq 0$, then $n^- \in \{n^-\}$. Therefore, $n^- \in n^- \cup \{n^-\} = n$. So, $n^- < n$.

Exercise 3.29: Determine if each of the following is true or false.

1. $\{\emptyset, \{\emptyset\}\} \in 2$ _____

2. $2 \in \big\{\emptyset, \{\emptyset\}, \{\emptyset, \{\emptyset\}\}\big\}$ _____

3. $4 <_\mathbb{N} \{0, 1, 2, 3\}$ _____

4. $17 \in 18$ _____

LEVEL 1

Determine if each of the following is true or false.

1. $(4, 5) = (5, 4)$

2. $\{4, 5\} = \{5, 4\}$

3. $(4, 5) = \{\{4\}, \{4, 5\}\}$

4. $(4, 4, 5) = (4, 5, 4)$

5. $\{4, 4, 5\} = \{4, 5\}$

6. $(4, 4, 5) = (4, 5)$

Use the roster method to describe each of the following sets.

7. $\{u\} \times \{v\} \times \{w\} \times \{z\} \times \{t\}$

8. $\{u, v, w\} \times \{z, t\}$

9. $\{z, t\}^2$

10. $\{z, t\}^3$

11. $\{z, t\}^4$

12. $\{u, v, w\}^2$

13. $\{u, v, w\}^3$

14. $\emptyset^2$

15. $\emptyset^3$

16. $\left(\mathcal{P}(\{\emptyset\}) \right)^2$

Let S be the set of people on a checkout line at a supermarket and define a relation $\prec$ on X by $x \prec y$ if x will check out before y.

17. Is $\prec$ reflexive on S?

18. Is $\prec$ symmetric on S?

19. Is $\prec$ transitive on S?

20. Is $\prec$ antireflexive on S?

21. Is $\prec$ antisymmetric on S?

22. Is $\prec$ trichotomous on S?

23. Does $\prec$ satisfy the comparability condition on S?

24. Is $(S, \prec)$ a poset?

25. Is $(S, \prec)$ a strict poset?

26. Is $(S, \prec)$ an ordered set?

Determine if each of the following is an ordered set.

27. $(\mathbb{Z}, >)$

28. $(\mathbb{Q}, <)$

29. $(\mathbb{Q}, \leq)$

30. $(\mathbb{R}, >)$

LEVEL 2

Find the domain, range, and field of each of the following relations:

31. $R = \{(a,b),(c,d),(e,f),(f,a)\}$

32. $S = \{(2k, 2t+1) \mid k, t \in \mathbb{Z}\}$

33. $T = \left\{(a,b) \mid \frac{a}{b} \in \mathbb{Q}\right\}$

34. $U = \mathbb{R} \times \mathbb{Z}^+$

Determine a and b so that each of the following equations is true, or state that the equation has no solution.

35. $\{a, b\} = \{8\}$

36. $\{a, b\} = \{7, 9\}$

37. $(a, b) = (7, 9)$

38. $(a, b) = \{\{7\}, \{9, 7\}\}$

39. $(a, b) = \{5\}$

40. $(a, b) = \{\{5\}\}$

41. $(a, a, b) = (1, b, a)$

42. $(a, b, c, d) = (1, c, 2, a)$

Suppose that $|A| = 5, |B| = 10, |C| = 50$, and $|D| = 100$. Compute each of the following:

43. $|A \times B|$

44. $|A \times B \times C \times D|$

45. $|C^2|$

46. $\mathcal{P}(C \times D)$

47. $\{R \mid R \text{ is a relation on } B\}$

48. $\{R \mid R \text{ is a relation on } B \times C\}$

LEVEL 3

Write each of the following in its unabbreviated form:

49. (x, x, z)

50. (x, z, x)

51. (x, x, x, x)

52. (x, y, z, w)

Let $A = \{a, b\}$ with $a \neq b$, consider the strict posets $(\mathbb{Z}, <)$ and $(\mathcal{P}(A), \subset)$, and let $<_D$ be the dictionary order on $\mathbb{Z} \times \mathcal{P}(A)$.

53. Is $<_D$ reflexive on $\mathbb{Z} \times \mathcal{P}(A)$?

54. Is $<_D$ symmetric on $\mathbb{Z} \times \mathcal{P}(A)$?

55. Is $<_D$ transitive on $\mathbb{Z} \times \mathcal{P}(A)$?

56. Is $<_D$ antireflexive on $\mathbb{Z} \times \mathcal{P}(A)$?

57. Is $<_D$ antisymmetric on $\mathbb{Z} \times \mathcal{P}(A)$?

58. Is $<_D$ trichotomous on $\mathbb{Z} \times \mathcal{P}(A)$?

59. Does $<_D$ satisfy the comparability condition on $\mathbb{Z} \times \mathcal{P}(A)$?

60. Is $(\mathbb{Z} \times \mathcal{P}(A), <_D)$ a poset?

61. Is $(\mathbb{Z} \times \mathcal{P}(A), <_D)$ a strict poset?

62. Is $(\mathbb{Z} \times \mathcal{P}(A), <_D)$ an ordered set?

Determine if each of the following is true or false.

63. $\{\emptyset, \{\emptyset\}\} \in 3$

64. $3 \in \{\emptyset, \{\emptyset\}, \{\emptyset, \{\emptyset\}\}\}$

65. $5 \in 8$

66. $5 <_{\mathbb{N}} \{0, 1, 2, 3, 4, 5, 6, 7\}$

LEVEL 4

Let A, B, C, and D be sets. Determine if each of the following statements is true or false. If true, explain why. If false, provide a counterexample.

67. $(A \times B) \cap (C \times D) = (A \cap C) \times (B \cap D)$.

68. $(A \times B) \cup (C \times D) = (A \cup C) \times (B \cup D)$.

69. If $A \subseteq B$ and $C \subseteq D$, then $A \times C \subseteq B \times D$.

Using the definition $(x, y) = \{\{x\}, \{x, y\}\}$, verify that each of the following is true.

70. If $x = z$ and $y = w$, then $(x, y) = (z, w)$.

71. If $x \neq y$ and $\{\{x\}, \{x, y\}\} = \{\{z\}, \{z, w\}\}$, then $x = z$.

72. If $x \neq y$ and $\{\{x\}, \{x, y\}\} = \{\{z\}, \{z, w\}\}$, then $y = w$.

73. If $x \neq y$ and $(x, y) = (z, w)$, then $x = z$ and $y = w$.

74. $\{\{x\}, \{x, x\}\} = \{\{x\}\}$.

75. If $(x, x) = (z, w)$, then $x = z = w$.

76. If $(x, y) = (z, w)$, then $x = z$ and $y = w$.

Verify that each of the following is true.

77. If $x = u$, $y = v$, and $z = w$, then $(x, y, z) = (u, v, w)$.

78. If $(x, y, z) = (u, v, w)$, then $x = u$, $y = v$, and $z = w$.

79. If $x_1 = y_1$, $x_2 = y_2$, $x_3 = y_3$, and $x_4 = y_4$, then $(x_1, x_2, x_3, x_4) = (y_1, y_2, y_3, y_4)$.

80. If $(x_1, x_2, x_3, x_4) = (y_1, y_2, y_3, y_4)$, then $x_1 = y_1$, $x_2 = y_2$, $x_3 = y_3$, and $x_4 = y_4$.

LEVEL 5

Let R be a relation on a set A. Determine if each of the following statements is true or false. If true, explain why. If false, provide a counterexample.

81. If R is symmetric and transitive on A, then R is reflexive on A.

82. If R is antisymmetric on A, then R is not symmetric on A.

For $a, b \in \mathbb{N}$, we will say that a divides b, written $a|b$, if there is a natural number k such that $b = ak$. Notice that | is a binary relation on $\mathbb{N}$.

83. Is | reflexive on $\mathbb{N}$?

84. Is | symmetric on $\mathbb{N}$?

85. Is | transitive on $\mathbb{N}$?

86. Is | antireflexive on $\mathbb{N}$?

87. Is | antisymmetric on $\mathbb{N}$?

88. Is | trichotomous on $\mathbb{N}$?

89. Does | satisfy the comparability condition on $\mathbb{N}$?

90. Is $(\mathbb{N}, |)$ a poset?

91. Is $(\mathbb{N}, |)$ a strict poset?

92. Is $(\mathbb{N}, |)$ an ordered set?

CHALLENGE PROBLEMS

93. Let $k, t, m \in \mathbb{N}$, let A and B be finite sets with $|A| = k$ and $|B| = t$, let $C = \left(\mathcal{P}(A \times B)\right)^m$, and let X be the set of relations on $A \times B \times C$. Evaluate $|X|$.

94. Let A, B, C, and D be sets. Explain why $\mathcal{P}\big((A \cap C) \times (B \cap D)\big) = \mathcal{P}(A \times B) \cap \mathcal{P}(C \times D)$.

Equivalence Relations

Recall from Lesson 3 that a binary relation R on a set A is

- **reflexive** if for all $a \in A$, aRa.

- **symmetric** if for all $a, b \in A$, aRb implies bRa.

- **transitive** if for all $a, b, c \in A$, aRb and bRc imply aRc.

A binary relation R on a set A is an **equivalence relation** if R is reflexive, symmetric, and transitive.

Example 4.1:

1. The most basic equivalence relation on a set A is the relation $R = \{(a, b) \in A^2 \mid a = b\}$ (the **equality relation**). We already saw in part 1 of Example 3.18 that this relation is reflexive, symmetric and transitive.

2. Another basic equivalence relation on a set A is the set A^2. Since every ordered pair (a, b) is in A^2, reflexivity, symmetry, and transitivity can never fail. We will refer to A^2 as the **trivial equivalence relation** on A.

   For example, if $A = \{a, b\}$, then $A^2 = \{(a, a), (a, b), (b, a), (b, b)\}$. Observe that this relation is reflexive, symmetric, and transitive.

3. Let $A = \{1, 2, 3\}$ and let $R = \{(1, 1), (2, 2), (3, 3), (1, 2), (2, 1)\}$. Then R is reflexive, symmetric, and transitive, and so, R is an equivalence relation.

Note: Remember that if R is a relation on a set A, then $(a, b) \in R$ is equivalent to aRb. So, in part 2 of Example 4.1, we have aRa, aRb, bRa, and bRb. Similarly, in part 3 of Example 4.1 above, we have $1R1$, $2R2$, $3R3$, $1R2$, and $2R1$.

Exercise 4.2: Let $A = \{0, 1, 2, 3\}$. Determine if each of the following relations is an equivalence relation on A.

1. $R = \{(0, 0), (1, 1), (2, 2), (3, 3)\}$ _____

2. $S = \{(0, 0), (1, 1), (2, 2), (0, 1), (1, 0)\}$ _____

3. $T = \{(0, 0), (1, 1), (2, 2), (3, 3), (0, 1), (1, 2), (0, 2)\}$ _____

4. $U = \{(0, 0), (1, 1), (2, 2), (3, 3), (0, 1), (1, 0), (1, 2), (2, 1), (0, 2), (2, 0)\}$ _____

5. $V = \{(0, 0), (1, 1), (2, 2), (3, 3), (0, 1), (1, 0), (1, 2), (2, 1)\}$ _____

Example 4.3: We say that integers a and b have the same **parity** if they are both even or both odd. Define $\equiv_2$ on $\mathbb{Z}$ by $\equiv_2 = \{(a, b) \in \mathbb{Z}^2 \mid a$ and b have the same parity$\}$.

1. $2 \equiv_2 4$ because 2 and 4 are both even. Similarly, $7 \equiv_2 15$ because 7 and 15 are both odd. On the other hand, $5 \not\equiv_2 10$ because 5 and 10 do not have the same parity (5 is odd and 10 is even).

2. $\equiv_2$ is reflexive on $\mathbb{Z}$ ($a \equiv_2 a$ because every integer has the same parity as itself).

3. $\equiv_2$ is symmetric on $\mathbb{Z}$ (if $a \equiv_2 b$, then a has the same parity as b, so b has the same parity as a, and therefore, $b \equiv_2 a$).

4. $\equiv_2$ is transitive on $\mathbb{Z}$ (if $a \equiv_2 b$ and $b \equiv_2 c$, then a, b, and c all have the same parity, and so, $a \equiv_2 c$).

5. Since $\equiv_2$ is reflexive, symmetric, and transitive on $\mathbb{Z}$, $\equiv_2$ is an equivalence relation on $\mathbb{Z}$.

Exercise 4.4: An integer n is **divisible** by 3, written $3 \mid n$, if there is another integer k such that $n = 3k$. Define $\equiv_3$ on $\mathbb{Z}$ by $\equiv_3 = \{(a, b) \in \mathbb{Z}^2 \mid 3 \mid b - a\}$.

1. Show that $\equiv_3$ is reflexive on $\mathbb{Z}$.

_____

_____

_____

2. Show that $\equiv_3$ is symmetric on $\mathbb{Z}$.

_____

_____

_____

3. Show that $\equiv_3$ is transitive on $\mathbb{Z}$.

_____

_____

_____

4. Is $\equiv_3$ an equivalence relation on $\mathbb{Z}$?

_____

_____

_____

Example 4.5: Consider the relation $R = \{((a,b),(c,d)) \in (\mathbb{N} \times \mathbb{N})^2 \mid a + d = b + c\}$ defined in part 5 of Example 3.14.

1. Since $a + b = b + a$, we see that $(a,b)R(a,b)$, and therefore, R is reflexive on $\mathbb{N} \times \mathbb{N}$.

2. If $(a,b)R(c,d)$, then $a + d = b + c$. Therefore, $c + b = d + a$, and so, $(c,d)R(a,b)$. Thus, R is symmetric on $\mathbb{N} \times \mathbb{N}$.

3. Suppose that $(a,b)R(c,d)$ and $(c,d)R(e,f)$. Then $a + d = b + c$ and $c + f = d + e$. So, $a + d + c + f = b + c + d + e$. Therefore, $a + f = b + e$, and so, we have $(a,b)R(e,f)$. So, R is transitive on $\mathbb{N} \times \mathbb{N}$.

4. Since R is reflexive, symmetric, and transitive on $\mathbb{N} \times \mathbb{N}$, it follows that R is an equivalence relation on $\mathbb{N} \times \mathbb{N}$.

Exercise 4.6: Consider the relation $R = \{((a,b),(c,d)) \in (\mathbb{Z} \times \mathbb{Z}^*)^2 \mid ad = bc\}$ defined in part 6 of Example 3.14.

1. Show that R is reflexive on $\mathbb{Z} \times \mathbb{Z}^*$.

_____

_____

_____

2. Show that R is symmetric on $\mathbb{Z} \times \mathbb{Z}^*$.

_____

_____

_____

3. Show that R is transitive on $\mathbb{Z} \times \mathbb{Z}^*$.

_____

_____

_____

4. Is R an equivalence relation on $\mathbb{Z} \times \mathbb{Z}^*$?

_____

_____

Let $\sim$ be an equivalence relation on a set S. If $x \in S$, the **equivalence class** of x, written $[x]$, is the set of all elements of S that are equivalent to x. Symbolically, we have

$$[x] = \{y \in S \mid x \sim y\}.$$

Example 4.7:

1. Let A be a set and let $R = \{(a, b) \in A^2 \mid a = b\}$ be the equality relation. Then for each $a \in A$, $[a] = \{a\}$ (because $a = a$ and no other element is equal to a).

2. Let A be a set and let $R = A^2$ be the trivial equivalence relation. Then for each $a \in A$, $[a] = A$ (because every element is equivalent to a).

3. Consider the equivalence relation $\equiv_2 = \{(a, b) \in \mathbb{Z}^2 \mid a \text{ and } b \text{ have the same parity}\}$ on $\mathbb{Z}$. Let $2\mathbb{Z} = \{2k \mid k \in \mathbb{Z}\}$ be the set of even integers and let $2\mathbb{Z} + 1 = \{2k + 1 \mid k \in \mathbb{Z}\}$ be the set of odd integers. We have $[0] = \{y \in \mathbb{Z} \mid 0 \equiv_2 y\} = 2\mathbb{Z}$. Observe that $[2] = [0]$, and in fact, if n is any even integer, then $[n] = [0] = 2\mathbb{Z}$. Similarly, if n is any odd integer, $[n] = [1] = 2\mathbb{Z} + 1$.

Exercise 4.8: Let $A = \{1, 2, 3\}$ and let $\sim = \{(1, 1), (2, 2), (3, 3), (1, 2), (2, 1)\}$. Determine if each of the following is true or false.

1. $\sim$ is an equivalence relation on A.

2. $1 \sim 2$ _____

3. $2 \sim 3$ _____

4. $1 \in [3]$ _____

5. $[1] = [2]$ _____

6. $[1] = [3]$ _____

7. $|[3]| = 2$ _____

8. $A = [1] \cup [3]$ _____

Partitions

Recall from Lesson 2 that sets A and B are **disjoint** or **mutually exclusive** if $A \cap B = \emptyset$.

Example 4.9:

1. Let $A = \{0, 1, 2\}$ and $B = \{3, 4, 5, 6\}$. Then $A \cap B = \emptyset$, and therefore, A and B are disjoint.

2. Let $A = \{x, y, z, w\}$ and let $B = \{u, v, w\}$. Then $A \cap B = \{w\}$, and therefore, A and B are **not** disjoint.

3. Let $2\mathbb{Z} = \{2k \mid k \in \mathbb{Z}\}$ be the set of even integers and let $2\mathbb{Z} + 1 = \{2k + 1 \mid k \in \mathbb{Z}\}$ be the set of odd integers. Then $2\mathbb{Z} \cap (2\mathbb{Z} + 1) = \emptyset$, and therefore, $2\mathbb{Z}$ and $2\mathbb{Z} + 1$ are disjoint.

If X is a nonempty set of sets, we say that X is **pairwise disjoint** if for all $A, B \in X$ with $A \neq B$, A and B are disjoint.

Example 4.10:

1. If A and B are disjoint, then $\{A, B\}$ is pairwise disjoint. For example, $\{\{0, 1, 2\}, \{3, 4, 5, 6\}\}$ is pairwise disjoint (see part 1 of Example 4.9 above) and $\{2\mathbb{Z}, 2\mathbb{Z} + 1\}$ is pairwise disjoint (see part 3 of Example 4.9 above).

2. Let $X = \{\{0, 1, 2\}, \{3, 4\}, \{5, 6, 7, 8\}\}$. Then X is pairwise disjoint because $\{0, 1, 2\} \cap \{3, 4\} = \emptyset$, $\{0, 1, 2\} \cap \{5, 6, 7, 8\} = \emptyset$, and $\{3, 4\} \cap \{5, 6, 7, 8\} = \emptyset$.

3. Let $X = \{\{0, 1, 2\}, \{2, 3\}, \{4, 5, 6, 7\}\}$. Then X is **not** pairwise disjoint. Indeed, we have $\{0, 1, 2\} \cap \{2, 3\} = \{2\} \neq \emptyset$. (However, note that $\cap X = \{0, 1, 2\} \cap \{2, 3\} \cap \{4, 5, 6, 7\} = \emptyset$. In this case, we say that X is **disjoint**. This example shows that a set of sets can be disjoint without being pairwise disjoint. We will not be interested in disjoint sets in this lesson.)

4. Let $3\mathbb{Z} = \{3k \mid k \in \mathbb{Z}\}$, $3\mathbb{Z} + 1 = \{3k + 1 \mid k \in \mathbb{Z}\}$, and $3\mathbb{Z} + 2 = \{3k + 2 \mid k \in \mathbb{Z}\}$. Then $X = \{3\mathbb{Z}, 3\mathbb{Z} + 1, 3\mathbb{Z} + 2\}$ is pairwise disjoint. We can visualize these three sets as follows:

$$3\mathbb{Z} = \{\ldots, -9, -6, -3, 0, 3, 6, 9, \ldots\}$$

$$3\mathbb{Z} + 1 = \{\ldots, -8, -5, -2, 1, 4, 7, 10, \ldots\}$$

$$3\mathbb{Z} + 2 = \{\ldots, -7, -4, -1, 2, 5, 8, 11, \ldots\}$$

5. If we let $3\mathbb{Z} + 3 = \{3k + 3 \mid k \in \mathbb{Z}\}$ and $X = \{3\mathbb{Z}, 3\mathbb{Z} + 1, 3\mathbb{Z} + 2, 3\mathbb{Z} + 3\}$, then X is **not** pairwise disjoint because $3\mathbb{Z} \cap (3\mathbb{Z} + 3) \neq \emptyset$. For example, $3 \in 3\mathbb{Z}$ because $3 = 3 \cdot 1$ and $3 \in 3\mathbb{Z} + 3$ because $3 = 3 \cdot 0 + 3$. In fact, it turns out that $3\mathbb{Z} = 3\mathbb{Z} + 3$.

Exercise 4.11: For each of the following, determine if X is pairwise disjoint.

1. $X = \{\{a, b\}, \{b, c\}, \{a, c\}\}$ _____

2. $X = \{\{a, b\}, \{c, d\}, \{e, a\}\}$ _____

3. $X = \{\{a, b\}, \{c, d\}, \{e, f\}\}$ _____

4. $X = \{2\mathbb{Z}, 3\mathbb{Z}\}$ _____

5. $X = \{2\mathbb{Z}, 3\mathbb{Z} + 1\}$ _____

6. $X = \{4\mathbb{Z}, 4\mathbb{Z} + 1, 4\mathbb{Z} + 2, 4\mathbb{Z} + 3\}$, where $4\mathbb{Z} = \{4k \mid k \in \mathbb{Z}\}$, $4\mathbb{Z} + 1 = \{4k + 1 \mid k \in \mathbb{Z}\}$, $4\mathbb{Z} + 2 = \{4k + 2 \mid k \in \mathbb{Z}\}$, and $4\mathbb{Z} + 3 = \{4k + 3 \mid k \in \mathbb{Z}\}$

   _____

7. $X = \{2\mathbb{Z}, 4\mathbb{Z} + 1\}$ _____

8. $X = \{2\mathbb{Z}, 4\mathbb{Z} + 1, 4\mathbb{Z} + 3\}$ _____

Recall from Lesson 2 that if X is a nonempty set of sets, then **union X** is defined as follows:

$$\cup X = \{y \mid \text{there is } Y \in X \text{ with } y \in Y\}.$$

See Example 2.14 and Exercise 2.15 for examples illustrating this definition.

A **partition** of a set S is a set of pairwise disjoint nonempty subsets of S whose union is S.

Example 4.12:

1. Let $A = \{0, 1, 2, 3, 4, 5\}$ and $X = \{\{0, 1, 5\}, \{2, 3, 4\}\}$. Then X is a partition of A because $\{0, 1, 5\} \cap \{2, 3, 4\} = \emptyset$ and $\cup X = \{0, 1, 5\} \cup \{2, 3, 4\} = \{0, 1, 2, 3, 4, 5\} = A$.

$Y = \{\{0, 1, 2, 5\}, \{2, 3, 4\}\}$ is **not** a partition of A because Y is not pairwise disjoint. Indeed, we have $\{0, 1, 2, 5\} \cap \{2, 3, 4\} = \{2\} \neq \emptyset$.

$Z = \{\{0, 1, 5\}, \{2, 3\}\}$ is **not** a partition of A because $\bigcup Z = \{0, 1, 5\} \cup \{2, 3\} = \{0, 1, 2, 3, 5\}$, and so, $\bigcup Z \neq A$ (we have $4 \in A \setminus \bigcup Z$).

2. Let $2\mathbb{Z} = \{2k \mid k \in \mathbb{Z}\}$ be the set of even integers and let $2\mathbb{Z} + 1 = \{2k + 1 \mid k \in \mathbb{Z}\}$ be the set of odd integers. Then $X = \{2\mathbb{Z}, 2\mathbb{Z} + 1\}$ is a partition of $\mathbb{Z}$. We can visualize this partition as follows:
$$\mathbb{Z} = \{\dots, -4, -2, 0, 2, 4, \dots\} \cup \{\dots, -3, -1, 1, 3, 5, \dots\}$$

3. For each $n \in \mathbb{N}$, let $A_n = \{2n, 2n + 1\}$. Then $X = \{A_n \mid n \in \mathbb{N}\}$ is a partition of $\mathbb{N}$. We can visualize this partition as follows:
$$\mathbb{N} = \{0, 1\} \cup \{2, 3\} \cup \{4, 5\} \cup \{6, 7\} \cup \{8, 9\} \cup \cdots$$

4. For each $n \in \mathbb{Z}$, let $A_n = \{(n, m) \mid m \in \mathbb{Z}\}$. Then $X = \{A_n \mid n \in \mathbb{Z}\}$ is a partition of $\mathbb{Z} \times \mathbb{Z}$. We can visualize this partition as follows:

$$\vdots \qquad \qquad \vdots \qquad \qquad \vdots$$

$$A_{-2} = \{\dots, (-2, -3), (-2, -2), (-2, -1), (-2, 0), (-2, 1), (-2, 2), (-2, 3), \dots\}$$
$$A_{-1} = \{\dots, (-1, -3), (-1, -2), (-1, -1), (-1, 0), (-1, 1), (-1, 2), (-1, 3), \dots\}$$
$$A_0 = \{\dots, (0, -3), (0, -2), (0, -1), (0, 0), (0, 1), (0, 2), (0, 3), \dots\}$$
$$A_1 = \{\dots, (1, -3), (1, -2), (1, -1), (1, 0), (1, 1), (1, 2), (1, 3), \dots\}$$
$$A_2 = \{\dots, (2, -3), (2, -2), (2, -1), (2, 0), (2, 1), (2, 2), (2, 3), \dots\}$$

$$\vdots \qquad \qquad \vdots \qquad \qquad \vdots$$

5. The only partition of the one element set $\{a\}$ is $\{\{a\}\}$. The partitions of the two element set $\{a, b\}$ with $a \neq b$ are $\{\{a\}, \{b\}\}$ and $\{\{a, b\}\}$.

Exercise 4.13: For each of the following, determine if X is a partition of A.

1. $A = \{a, b, c\}$; $X = \{\{a\}, \{b\}, \{c\}\}$ _____

2. $A = \{a, b, c\}$; $X = \{\{a, b\}, \{b, c\}\}$ _____

3. $A = \{a, b, c\}$; $X = \{\{a, b\}, \{c\}\}$ _____

4. $A = \mathbb{Z}$; $X = \{3\mathbb{Z}, 3\mathbb{Z} + 1, 3\mathbb{Z} + 2\}$ _____

5. $A = \mathbb{C}$; $X = \{X_a \mid a \in \mathbb{R}\}$, where for each $a \in \mathbb{R}$, $X_a = \{a + bi \mid b \in \mathbb{R}\}$

_____

The following basic facts provide a useful relationship between equivalence relations (ERs) and partitions.

ER-Partition Fact 1: If P is a partition of a set S, then there is an equivalence relation $\sim$ on S for which the elements of P are the equivalence classes of $\sim$.

ER-Partition Fact 2: If $\sim$ is an equivalence relation on a set S, then the equivalence classes of $\sim$ form a partition of S.

You will be asked to verify these facts in Problems 66 – 77 below.

Example 4.14: Consider the equivalence relation $\equiv_2$ from Example 4.3, defined by $a \equiv_2 b$ if and only if a and b have the same parity, and the partition $\{2\mathbb{Z}, 2\mathbb{Z} + 1\}$ of $\mathbb{Z}$ from part 2 of Example 4.12. For this partition, we are thinking of $\mathbb{Z}$ as the union of the even and odd integers:

$$\mathbb{Z} = \{\dots, -4, -2, 0, 2, 4, \dots\} \cup \{\dots, -3, -1, 1, 3, 5, \dots\}$$

Observe that a and b are in the same member of the partition if and only if $a \equiv_2 b$ if and only if $[a] = [b]$. If n is any even integer, then we have $[n] = [0] = 2\mathbb{Z}$ and if n is any odd integer, then we have $[n] = [1] = 2\mathbb{Z} + 1$.

Example 4.15: Recall from Lesson 1 that the **power set** of A, written $\mathcal{P}(A)$, is the set consisting of all subsets of A.

$$\mathcal{P}(A) = \{X \mid X \subseteq A\}$$

For example, if $A = \{a, b, c\}$, then $\mathcal{P}(A) = \{\emptyset, \{a\}, \{b\}, \{c\}, \{a, b\}, \{a, c\}, \{b, c\}, \{a, b, c\}\}$. We can define a binary relation $\sim$ on $\mathcal{P}(A)$ by $X \sim Y$ if and only if $|X| = |Y|$ (X and Y have the same number of elements). Then $\sim$ is an equivalence relation on $\mathcal{P}(A)$. There are four equivalence classes.

$$[\emptyset] = \{\emptyset\} \qquad\qquad [\{a\}] = \{\{a\}, \{b\}, \{c\}\}$$
$$[\{a, b\}] = \{\{a, b\}, \{a, c\}, \{b, c\}\} \qquad\qquad [\{a, b, c\}] = \{\{a, b, c\}\}$$

Notes: (1) $\{a\} \sim \{b\} \sim \{c\}$ because each of these sets has one element. It follows that $\{a\}$, $\{b\}$, and $\{c\}$ are all in the same equivalence class. Above, we chose to use $\{a\}$ as the **representative** for this equivalence class. This is an arbitrary choice. In fact, $[\{a\}] = [\{b\}] = [\{c\}]$.

Similarly, $[\{a, b\}] = [\{a, c\}] = [\{b, c\}]$.

(2) The empty set is the only subset of A with 0 elements. Therefore, the equivalence class of $\emptyset$ contains only itself. Similarly, the equivalence class of $A = \{a, b, c\}$ contains only itself.

(3) Notice that the four equivalence classes are pairwise disjoint, nonempty, and their union is $\mathcal{P}(A)$. In other words, the equivalence classes form a partition of $\mathcal{P}(A)$.

Important note: We will sometimes want to define relations on equivalence classes. When we do this, we must be careful that what we are defining is **well-defined**.

For example, consider the equivalence relation $\equiv_2$ on $\mathbb{Z}$, and let $X = \{[0], [1]\}$ be the set of equivalence classes.

Let's attempt to define a relation on X by $[x]R[y]$ if and only if $x < y$. Is $[0]R[1]$ true? It looks like it is because $0 < 1$. But this isn't the end of the story. Since $[0] = [2]$, if $[0]R[1]$, then we must also have $[2]R[1]$ (by a direct substitution). But $2 \not< 1$! So, $[2]R[1]$ is false. To summarize, $[0]R[1]$ should be true and $[2]R[1]$ should be false, but $[0]R[1]$ and $[2]R[1]$ represent the same statement. So, R is **not** a well-defined relation on X.

The Integers

At this point, let's provide a more formal definition of the set of integers.

To motivate the definition of the integers, note that we can think of every integer as a difference of two natural numbers. For example, the integer -3 can be thought of as $1-4$. However, -3 can also be thought of as $2-5$. So, we must insist that $1-4=2-5$, or equivalently, $1+5=2+4$.

We define a relation R on $\mathbb{N} \times \mathbb{N}$ by $R = \{((a,b),(c,d)) \in (\mathbb{N} \times \mathbb{N})^2 \mid a+d=b+c\}$. In Example 4.5, we showed that this relation is an equivalence relation. We can now define the set of integers to be the set of equivalence classes for this equivalence relation. That is, we define the set of integers to be $\mathbb{Z} = \{[(a,b)] \mid (a,b) \in \mathbb{N} \times \mathbb{N}\}$.

Example 4.16:

1. $[(k,k)] = [(0,0)]$ for all $k \in \mathbb{N}$ because $k+0 = k+0$. We identify the integer $[(0,0)]$ with the natural number 0.

2. $[(5,0)] = [(6,1)]$ because $5+1=0+6$. Similarly, we have $[(5,0)] = [(7,2)] = [(8,3)]$, and in general $[(5,0)] = [(5+k,k)]$ for any natural number k. $[(5,0)]$ is the most "natural" way to express the natural number 5 as an integer. More generally, the natural number n can be expressed as an integer as $[(n,0)]$.

3. We usually abbreviate the integer $[(0,k)]$ as $-k$. For example, -3 is an abbreviation for $[(0,3)]$. We can also write -3 as $[(1,4)]$ because $1+3=4+0$.

4. If $a,b \in \mathbb{N}$ with $a \geq b$, then $[(a,b)] = [(a-b,0)]$. If $a < b$, then $[(a,b)] = [(0,b-a)]$. In this way, we see that every integer can be written in the form $[(n,0)]$ or $[(0,n)]$ for some $n \in \mathbb{N}$. We abbreviate $[(n,0)]$ by n and we abbreviate $[(0,n)]$ by $-n$. For example, we have $[(2,7)] = [(0,5)] = -5$.

Exercise 4.17: Determine if each of the following is true or false.

1. $[(36,36)] = 36$ _____

2. $[(5,5)] = 0$ _____

3. $[(1,2)] = [(2,1)]$ _____

4. $[(3,4)] = -1$ _____

5. $[(4,3)] = -1$ _____

We define the ordering $<_\mathbb{Z}$ on $\mathbb{Z}$ by $[(a,b)] <_\mathbb{Z} [(c,d)]$ if and only if $a+d <_\mathbb{N} b+c$, where $<_\mathbb{N}$ is the usual ordering on $\mathbb{N}$ ($n <_\mathbb{N} m$ if and only if $n \in m$).

Note: We will usually abbreviate $<_\mathbb{Z}$ simply by $<$, especially if it is already clear that we are working with the integers.

Example 4.18:

1. $[(0,0)] < [(4,0)]$ because $0 + 0 < 0 + 4$. More generally, for any natural number $k \neq 0$, we have $[(0,0)] < [(k,0)]$ because $0 + 0 < 0 + k$. This shows that for any natural number $k \neq 0$, the natural number k satisfies $0 < k$.

2. $[(0,4)] < [(0,0)]$ because $0 + 0 < 4 + 0$. More generally, for any natural number $k \neq 0$, we have $[(0,k)] < [(0,0)]$ because $0 + 0 < k + 0$. This shows that for any natural number $k \neq 0$, the integer $-k$ satisfies $-k < 0$.

Exercise 4.19: For each of the following, replace $\square$ by $<$, $>$, or $=$ to get a true statement.

1. $[(5,7)] \,\square\, [(2,11)]$ _____

2. $[(26,5)] \,\square\, [(25,3)]$ _____

3. $[(5,17)] \,\square\, [(10,22)]$ _____

4. $[(3,4)] \,\square\, [(4,3)]$ _____

In Problems 58 – 61 below, you will be asked to show that $<_{\mathbb{Z}}$ is a well-defined strict linear ordering on $\mathbb{Z}$.

The Rational Numbers

Now, let's provide a more formal definition of the set of rational numbers.

We define a relation R on $\mathbb{Z} \times \mathbb{Z}^*$ by $R = \{((a,b),(c,d)) \in (\mathbb{Z} \times \mathbb{Z}^*)^2 \mid ad = bc\}$. In Exercise 4.6, you were asked to show that this relation is an equivalence relation.

For each $a \in \mathbb{Z}$ and $b \in \mathbb{Z}^*$, we define the **rational number** $\frac{a}{b}$ to be the equivalence class of (a,b). So, $\frac{a}{b} = [(a,b)]$, and we have $\frac{a}{b} = \frac{c}{d}$ if and only if $(a,b)R(c,d)$ if and only if $ad = bc$.

The set of rational numbers is $\mathbb{Q} = \left\{ \frac{a}{b} \mid a \in \mathbb{Z} \wedge b \in \mathbb{Z}^* \right\}$. In words, $\mathbb{Q}$ is "the set of quotients a over b such that a and b are integers and b is not zero."

Example 4.20:

1. $\frac{1}{3} = \frac{2}{6}$ (or equivalently, $[(1,3)] = [(2,6)]$) because $1 \cdot 6 = 3 \cdot 2$.

2. $\frac{-5}{-3} = \frac{5}{3}$ because $(-5)(3) = (-3)(5)$.

3. We identify the rational number $\frac{a}{1}$ with the integer a. In this way, we have $\mathbb{Z} \subseteq \mathbb{Q}$. For example, $\frac{5}{1} = 5$.

Exercise 4.21: Determine if each of the following is true or false.

1. $\frac{3}{3} = 1$ _____

2. $\frac{1}{7} = \frac{7}{1}$ _____

3. $\frac{-2}{7} = \frac{4}{-14}$ _____

We define $<_\mathbb{Q}$ on $\mathbb{Q}$ by $\frac{a}{b} <_\mathbb{Q} \frac{c}{d}$ if and only if $ad <_\mathbb{Z} bc$, where $<_\mathbb{Z}$ is the usual ordering on $\mathbb{Z}$.

Note: We will usually abbreviate $<_\mathbb{Q}$ simply by $<$, especially if it is already clear that we are working with the rational numbers.

In Problem 78 below, you will be asked to show that $<_\mathbb{Q}$ is a well-defined strict linear ordering on $\mathbb{Q}$.

Example 4.22:

1. $\frac{2}{3} < \frac{5}{4}$ because $2 \cdot 4 < 3 \cdot 5$.

2. $\frac{-3}{4} < \frac{-5}{7}$ because $-3 \cdot 7 < 4(-5)$.

Problem Set 4

Full solutions to these problems are available for free download here:
www.SATPrepGet800.com/STKZ3D

LEVEL 1

Let $X = \{a, b, c\}$. Determine if each of the following is an equivalence relation on X.

1. $\{(a, a), (b, b), (c, c)\}$

2. $\{(a, a), (a, b), (a, c)\}$

3. $\{(a, a), (b, a), (c, a)\}$

4. $\{(a, a), (a, b), (a, c), (b, a), (b, b), (b, c), (c, a), (c, b), (c, c)\}$

5. $\{(a, a), (a, b), (a, c), (b, b), (b, c), (c, c)\}$

6. $\{(a, a), (a, b), (b, b), (b, a), (c, c)\}$

7. $\{(a, a), (a, b), (b, a), (b, b), (b, c), (c, b), (c, c)\}$

8. $\{(a, a), (b, b), (b, c), (c, c), (c, b)\}$

Let $X = \{0, 1, 2, 3, 4\}$. For each of the following, describe a relation R on X with the given properties:

9. R is reflexive, but not symmetric.

10. R is reflexive, but not transitive.

11. R is symmetric, but not reflexive.

12. R is symmetric, but not transitive.

13. R is transitive, but not reflexive.

14. R is transitive, but not symmetric.

Find all partitions of each of the following sets:

15. $\{x\}$

16. $\{x, y\}$

17. $\{x, y, z\}$

18. $\{x, y, z, w\}$

LEVEL 2

For each of the following equivalence relations on $A = \{a, b, c, d\}$, find $[a]$, $[b]$, $[c]$, and $[d]$.

19. $\{(a, a), (b, b), (c, c), (d, d)\}$

20. $\{(a, a), (a, b), (b, a), (b, b), (c, c), (d, d)\}$

21. $\{(a, a), (a, b), (a, c), (b, a), (b, b), (b, c), (c, a), (c, b), (c, c), (d, d)\}$

22. $\{(a, a), (a, c), (b, b), (b, d), (c, a), (c, c), (d, b), (d, d)\}$

Find an equivalence relation R on $A = \{0, 1, 2\}$ with the given property.

23. If S is any equivalence relation on A, then $R \subseteq S$.

24. If S is any equivalence relation on A, then $S \subseteq R$.

25. $|R| = 5$ and $0R1$.

Determine if each of the following is a partition of $\{0, 1, 2, 3, 4, 5, 6, 7, 8\}$.

26. $\{\{0\}, \{1\}, \{2\}, \{3\}, \{4\}, \{5\}, \{6\}, \{7\}, \{8\}\}$

27. $\{\{0, 1\}, \{1, 2, 3\}, \{3, 4, 5, 6\}, \{6, 7, 8\}\}$

28. $\{\{0, 1, 2, 3, 4, 5, 6, 7, 8\}\}$

29. $\{\{0, 1, 2, 3\}, \{5, 6, 7, 8\}\}$

30. $\{\{0, 2, 4, 6, 8\}, \{1, 3, 5, 7\}\}$

Let $X = \{0, 1, 2, 3, 4\}$. For each of the following, find the equivalence relation R on X of least cardinality containing the given elements.

31. $(0, 1) \in R$.

32. $(0, 1), (2, 3) \in R$.

33. $(0, 1), (1, 2), (2, 3) \in R$.

34. $(0, 1), (2, 1), (3, 4), (4, 2) \in R$.

Determine if each of the following is a partition of $\mathbb{Z}$.

35. $\{\mathbb{Z}\}$

36. $\{3\mathbb{Z}, 3\mathbb{Z} + 1\}$

37. $\{\mathbb{Z}^-, \mathbb{Z}^+\}$

38. $\{\mathbb{N}, \{-n \mid n \in \mathbb{N}\}\}$

39. $\{5\mathbb{Z}, 5\mathbb{Z} + 1, 5\mathbb{Z} + 2, 5\mathbb{Z} + 3, 5\mathbb{Z} + 4\}$

Let $R = \{((a, b), (c, d)) \in (\mathbb{N} \times \mathbb{N})^2 \mid a + d = b + c\}$, so that $\mathbb{Z} = \{[(a, b)] \mid (a, b) \in \mathbb{N} \times \mathbb{N}\}$. Determine if each of the following is true or false.

40. $[(11, 12)] = [(12, 11)]$

41. $[(9, 9)] = 0$

42. $[(0, 1)] = 1$

43. $[(9, 7)] < [(7, 9)]$

44. $[(0, 2)] < [(3, 5)]$

45. $[(4, 4)] > [(3, 4)]$

LEVEL 4

For $n, m \in \mathbb{Z}$, we will say that n **divides** m, written $n|m$, if there is an integer k such that $m = nk$. For $n \in \mathbb{Z}^+$, define $\equiv_n$ on $\mathbb{Z}$ by $\equiv_n = \{(a, b) \in \mathbb{Z}^2 \mid n|b - a\}$.

46. Show that $\equiv_n$ is reflexive on $\mathbb{Z}$.

47. Show that $\equiv_n$ is symmetric on $\mathbb{Z}$.

48. Show that $\equiv_n$ is transitive on $\mathbb{Z}$.

49. Is $\equiv_n$ an equivalence relation on $\mathbb{Z}$?

Let X be a set of equivalence relations on a set A. For each of the following statements, either verify that the statement is true or provide a counterexample showing that the statement is false.

50. $\cap X$ is reflexive on A.

51. $\cup X$ is reflexive on A.

52. $\cap X$ is symmetric on A.

53. $\cup X$ is symmetric on A.

54. $\cap X$ is transitive on A.

55. $\cup X$ is transitive on A.

56. $\cap X$ is an equivalence relation on A.

57. $\cup X$ is an equivalence relation on A.

LEVEL 5

Recall that we define the ordering $<_{\mathbb{Z}}$ on $\mathbb{Z}$ by $[(a, b)] <_{\mathbb{Z}} [(c, d)]$ if and only if $a + d <_{\mathbb{N}} b + c$, where $<_{\mathbb{N}}$ is the usual ordering on $\mathbb{N}$.

58. Show that $<_{\mathbb{Z}}$ is a well-defined relation on $\mathbb{Z}$.

59. Show that $<_{\mathbb{Z}}$ is transitive on $\mathbb{Z}$.

60. Show that $<_{\mathbb{Z}}$ is trichotomous on $\mathbb{Z}$.

61. Explain why $(\mathbb{Z}, <_{\mathbb{Z}})$ is an ordered set.

Let $R = \{(x, y) \in \mathbb{R} \times \mathbb{R} \mid x - y \in \mathbb{Z}\}$.

62. Show that R is reflexive on $\mathbb{R}$.

63. Show that R is symmetric on $\mathbb{R}$.

64. Show that R is transitive on $\mathbb{R}$.

65. Is R an equivalence relation on $\mathbb{R}$?

Let $\sim$ be an equivalence relation on a set S.

66. Show that $\bigcup\{[x] \mid x \in S\} \subseteq S$.

67. Show that $S \subseteq \bigcup\{[x] \mid x \in S\}$.

68. Suppose that $x, y \in S$ and $[x] \cap [y] \neq \emptyset$. Show that $[x] \subseteq [y]$.

69. Suppose that $x, y \in S$ and $[x] \cap [y] \neq \emptyset$. Explain why $[x] = [y]$.

70. Show that the equivalence classes of $\sim$ form a partition of S.

Let P be a partition of S, and define the relation $\sim$ by $x \sim y$ if and only if there is $X \in P$ with $x, y \in X$.

71. Show that $\sim$ is reflexive on S.

72. Show that $\sim$ is symmetric on S.

73. Show that $\sim$ is transitive on S.

74. Show that $\sim$ is an equivalence relation on S.

75. Show that $P \subseteq \{[x] \mid x \in S\}$.

76. Show that $\{[x] \mid x \in S\} \subseteq P$.

77. Show that $P = \{[x] \mid x \in S\}$.

78. Recall that we define the ordering $<_{\mathbb{Q}}$ on $\mathbb{Q}$ by $\frac{a}{b} <_{\mathbb{Q}} \frac{c}{d}$ if and only if $ad <_{\mathbb{Z}} bc$, where $<_{\mathbb{Z}}$ is the usual ordering on $\mathbb{Z}$. Show that $<_{\mathbb{Q}}$ is a well-defined strict linear ordering on $\mathbb{Q}$.

79. Let R and S be binary relations on a set A. The composition of R and S, written $R \circ S$, is defined as $R \circ S = \{(a, b) \mid \exists c \in A((a, c) \in R \wedge (c, b) \in S)\}$. Suppose that R and S are equivalence relations on A. Show that $R \circ S$ is an equivalence relation on A if and only if $R \circ S = S \circ R$.

80. Define a partition P of $\mathbb{N}$ such that $P \sim \mathbb{N}$ and for each $X \in P$, $X \sim \mathbb{N}$.

LESSON 5
FUNCTIONS

Function Basics

Let A and B be sets. Informally, a **function** from A to B is a rule that assigns to each element in A exactly one element in B.

Example 5.1: Let $A = \{\text{dog, shark, hippo, penguin, snake}\}$ and let $B = \mathbb{N}$. We can define a function from A to B by assigning to each animal in A the number of legs the animal has. Under this function, dog and hippo are assigned the natural number 4, shark and snake are assigned the natural number 0, and penguin is assigned the natural number 2. We can visualize this function as follows:

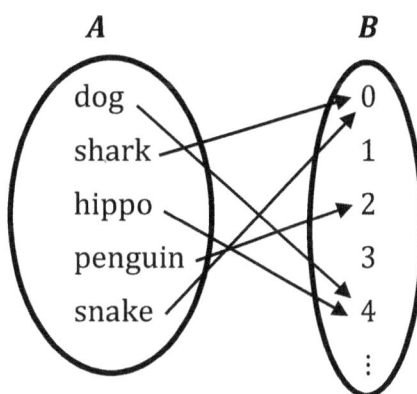

Let's now provide a more formal definition of a function.

Formally, f is a **function** from A to B, written $f: A \to B$, if the following two conditions hold.

1. $f \subseteq A \times B$.

2. For all $a \in A$, there is a unique $b \in B$ such that $(a, b) \in f$.

Example 5.2:

1. In Example 5.1, we defined a function from $A = \{\text{dog, shark, hippo, penguin, snake}\}$ to $B = \mathbb{N}$ by assigning each animal in A to the number of legs the animal has. If we name this function f, then we see that $f = \{(\text{dog}, 4), (\text{shark}, 0), (\text{hippo}, 4), (\text{penguin}, 2), (\text{snake}, 0)\}$. Notice that f meets both requirements for being a function: f is a subset of $A \times B$ and for each $a \in A$, there is a unique $b \in B$ such that $(a, b) \in f$. So, we may write $f: A \to B$.

2. Let A and B be as in part 1 above, and let $g = \{(\text{dog}, 4), (\text{shark}, 0), (\text{hippo}, 4), (\text{penguin}, 2)\}$. Then $g \subseteq A \times B$ (the first requirement is satisfied). Now, snake $\in A$, but there is no natural number b such that $(\text{snake}, b) \in g$. Therefore, requirement 2 is **not** satisfied. So, g is not a function from A to B.

   However, in this case, g is still a function. Indeed, if we replace the set A by the set $C = \{\text{dog, shark, hippo, penguin}\}$, then we see that g is a function from C to B. Symbolically, we have $g: C \to B$, whereas $g: A \nrightarrow B$.

75

3. Now, let $h = \{(\text{dog}, 4), (\text{dog}, 5), (\text{shark}, 0), (\text{hippo}, 4), (\text{penguin}, 2), (\text{snake}, 0)\}$ (once again, A and B are as is part 1 above). Then h meets the first requirement for being a function: $h \subseteq A \times B$. However, the second requirement is violated: the "input" dog has the two distinct "outputs" 4 and 5.

   Unlike part 2 above, there is no simple way to "repair" the problem here. We would need to remove one of the ordered pairs $(\text{dog}, 4)$ or $(\text{dog}, 5)$ in order to be left with a function from A to B.

Exercise 5.3: Let $A = \{0, 1, 2, 3\}$ and let $B = \{a, b, c\}$. Determine if each of the following is a function from A to B.

1. $f = \{(0, a), (1, b), (2, c)\}$ _____

2. $g = \{(0, b), (1, a), (2, c), (3, b)\}$ _____

3. $h = \{(0, b), (1, c), (2, b), (2, c), (3, a)\}$ _____

4. $k = \{(0, c), (1, c), (2, c), (3, c)\}$ _____

5. $m = \{(a, 0), (b, 1), (c, 2)\}$ _____

Notes: (1) A function $f: A \to B$ is a binary relation on $A \cup B$.

(2) Not every binary relation on $A \cup B$ is a function from A to B. See parts 2 and 3 of Example 5.2 above for simple counterexamples.

(3) The uniqueness in the second clause in the definition of a function is equivalent to the statement "if $(a, b), (a, c) \in f$, then $b = c$."

(4) When we know that f is a function, we will abbreviate $(a, b) \in f$ by $f(a) = b$.

If $f: A \to B$, the **domain** of f, written dom f, is the set A, and the **range** of f, written ran f, is the set $\{f(a) \mid a \in A\}$. Observe that ran $f \subseteq B$. The set B itself is sometimes called the **codomain** of f.

Example 5.4:

1. $f = \{(0, a), (1, a)\}$ is a function with dom $f = \{0, 1\}$ and ran $f = \{a\}$. Instead of $(0, a) \in f$, we will usually write $f(0) = a$. Similarly, instead of $(1, a) \in f$, we will write $f(1) = a$. Here is a visual representation of this function.

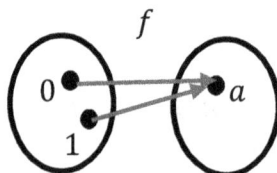

   This function $f: \{0, 1\} \to \{a\}$ is called a **constant function** because the range of f consists of a single element.

   Note also that f is a binary relation on the set $\{0, 1, a\}$.

2. If $a \neq b$, then $g = \{(0, a), (0, b)\}$ is **not** a function because it violates the second clause in the definition of being a function. It is, however, a binary relation on $\{0, a, b\}$ with dom $g = \{0\}$ and ran $g = \{a, b\}$.

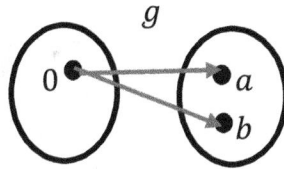

3. $h = \{(a, b) \mid a, b \in \mathbb{R} \wedge a > 0 \wedge a^2 + b^2 = 2\}$ is a relation on $\mathbb{R}$ that is **not** a function. $(1, 1)$ and $(1, -1)$ are both elements of h, violating the second requirement in the definition of a function. See the figure below on the left. Notice how a vertical line hits the graph twice.

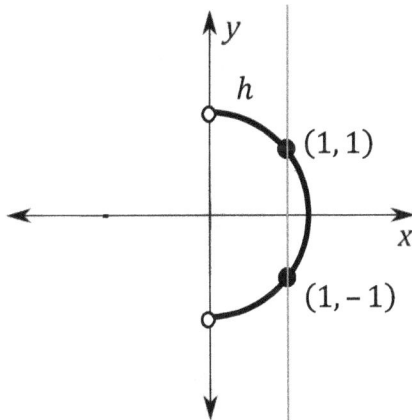

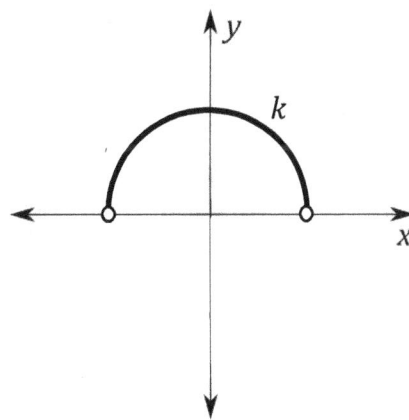

4. $k = \{(a, b) \mid a, b \in \mathbb{R} \wedge b > 0 \wedge a^2 + b^2 = 2\}$ **is** a function. See the figure above on the right. To see that the second clause in the definition of a function is satisfied, suppose that (a, b) and (a, c) are both in f. Then $a^2 + b^2 = 2$, $a^2 + c^2 = 2$, and b and c are both positive. It follows that $b^2 = c^2$, and since b and c are both positive, we have $b = c$.

We have dom $k = \left(-\sqrt{2}, \sqrt{2}\right)$ and ran $k = \left(0, \sqrt{2}\,\right]$. So, $k: \left(-\sqrt{2}, \sqrt{2}\right) \to \left(0, \sqrt{2}\,\right]$.

Note that if $a \in \mathbb{R}$, then $a^2 = a \cdot a$. Also, if $a \geq 0$ and $a^2 = b$, then $\sqrt{b} = a$.

5. $b = \{(z, w) \mid z, w \in \mathbb{C} \wedge w = z - 1\}$ is a function. By Note 4 above, we can describe b using the notation $b(z) = z - 1$.

If we write $z = x + yi$ and $b(z) = u + vi$, then $b(x + yi) = x + yi - 1 = (x - 1) + yi$ and we see that $u(x, y) = x - 1$ and $v(x, y) = y$.

b is an example of a complex-valued function. One way to visualize this function is to simply stay in the same plane and to analyze how a typical point moves or how a certain set is transformed. The function b takes the point (x, y) to the point $(x - 1, y)$. That is, each point is shifted one unit to the left. Similarly, if $S \subseteq \mathbb{C}$, then each point of the set S is shifted one unit to the left by the function b. Both these situations are demonstrated in the figure to the right. Observe how the point $(2, 2)$ is shifted to the point $(1, 2)$ and how each point of the rightmost rectangle is shifted one point to the left to form the leftmost rectangle.

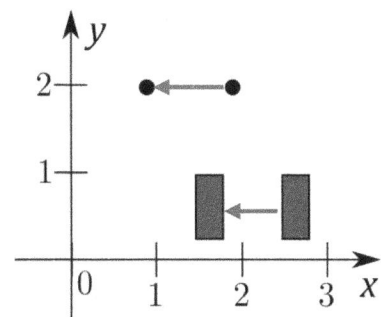

A second way to visualize this function is to draw two separate planes: an xy-plane and a uv-plane. We can then draw a point or a set in the xy-plane and its image under b in the uv-plane. In the figure below, we do this for the same point and the same rectangle as we did in the previous figure.

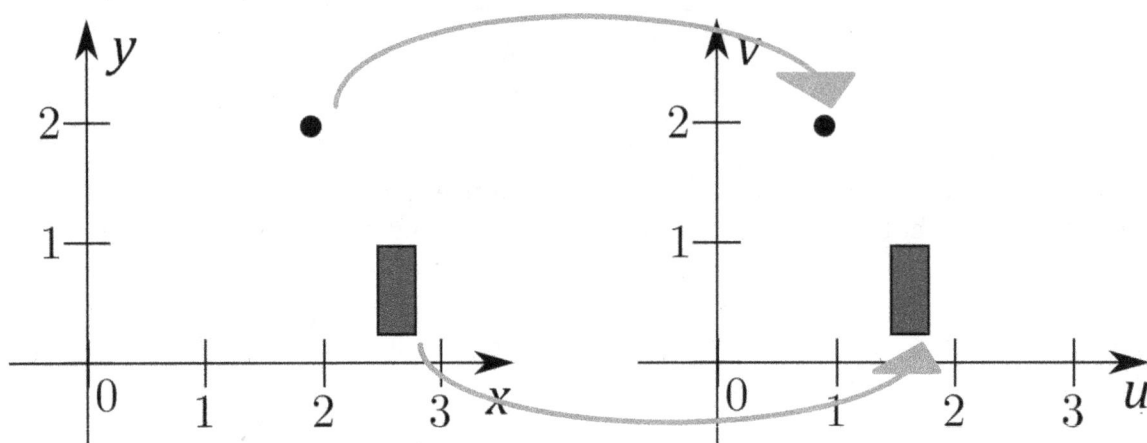

Exercise 5.5: Let $A = \{a, b\}$ and let $B = \{0, 1, 2\}$.

1. List all the functions from A to B.

_____   _____   _____

_____   _____   _____

_____   _____   _____

2. List all the functions from B to A.

_____   _____

_____   _____

Sequences

An **infinite sequence** is a function with domain $\mathbb{N}$. A nice way to visualize an infinite sequence is to list the "outputs" of the sequence in order in parentheses. So, if A is a nonempty set and $f: \mathbb{N} \to A$ is a sequence, then we can express f as $(f(0), f(1), f(2), \dots)$.

Example 5.6: Let $f: \mathbb{N} \to \{0, 1\}$ be defined by $f(n) = \begin{cases} 0 & \text{if } n \text{ is even.} \\ 1 & \text{if } n \text{ is odd.} \end{cases}$ This is an **explicit definition** of the function f. We can visualize this sequence f as the **infinite tuple** $(0, 1, 0, 1, 0, 1, \dots)$. We can also express this sequence as the following **set of ordered pairs**:

$$\{(0, 0), (1, 1), (2, 0), (3, 1), (4, 0), (5, 1), \dots\}$$

Summary: We have just seen the following three ways to describe a sequence.

- An explicit definition
- An infinite tuple
- A set of ordered pairs

Exercise 5.7: Each of the following sequences is described in one of the three ways specified in the summary above. In each case, describe the sequence in the other two ways.

1. $f: \mathbb{N} \to \mathbb{Q}$, where $f(n) = \frac{n}{2}$ _____ _____

2. $(1, 2, 3, 4, 5, 6, 7, \ldots)$ _____ _____

3. $\{(0, 0), (1, 5), (2, 10), (3, 15), (4, 20), (5, 25), (6, 30), \ldots\}$

_____ _____

Informally, it may be most natural to think of an infinite sequence as an infinite length tuple. If $f: \mathbb{N} \to A$ is defined by $f(n) = a_n$, then we may represent the sequence f using the notation $(a_n)_{n \in \mathbb{N}}$ or simply (a_n). When we use this notation, a_n is called the nth term of the sequence.

Example 5.8: $(1, 2, 4, 8, 16, 32, \ldots)$ represents the sequence $g: \mathbb{N} \to \mathbb{N}$ defined by $g(n) = 2^n$. This sequence g can be written as $(2^n)_{n \in \mathbb{N}}$ or more simply as (2^n). The 0th term of the sequence g is 1, the first term of the sequence g is 2, and so on. In general, the nth term of the sequence g is 2^n.

Exercise 5.9: Express each of the three sequences from Exercise 5.7 using the notation (a_n). What is the nth term of each of these sequences?

1. $f: \mathbb{N} \to \mathbb{Q}$, where $f(n) = \frac{n}{2}$ _____ _____

2. $(1, 2, 3, 4, 5, 6, 7, \ldots)$ _____ _____

3. $\{(0, 0), (1, 5), (2, 10), (3, 15), (4, 20), (5, 25), (6, 30), \ldots\}$

_____ _____

A **finite sequence of length n** is a function with domain $n = \{0, 1, \ldots, n - 1\}$. We can visualize such a sequence as an n-tuple or a set of ordered pairs similarly to the way we can visualize an infinite sequence.

Recall: We define the natural numbers by letting $0 = \emptyset$, $1 = \{0\}$, $2 = \{0, 1\}$, $3 = \{0, 1, 2\}$,... and so on. In general, the natural number n is the set of all its predecessors. Specifically, $n = \{0, 1, 2, \ldots, n - 1\}$. Using this notation, we can say that a finite sequence of length n is a function $f: n \to A$ for some set A.

For example, a sequence of length 3 has domain $3 = \{0, 1, 2\}$ and if $f: 3 \to A$ is such a sequence, then we can represent the sequence as the ordered 3-tuple (or ordered triple) $(f(0), f(1), f(2))$ or the set of ordered pairs $\{(0, f(0)), (1, f(1)), (2, f(2))\}$.

Example 5.10: The sequence $(0, 2, 4, 6, 8, 10)$ is the function $h: 6 \to \mathbb{N}$ (or equivalently, $h: \{0, 1, 2, 3, 4, 5\} \to \mathbb{N}$) defined by $h(k) = 2k$. This sequence has length 6. We can also express this sequence as the following **set of ordered pairs**:

$$\{(0, 0), (1, 2), (2, 4), (3, 6), (4, 8), (5, 10)\}$$

Observe how a finite sequence with domain $n = \{0, 1, \ldots, n-1\}$ and range A looks just like an n-tuple in A^n. In fact, it's completely natural to identify a finite sequence of length n with the corresponding n-tuple. So, $(0, 2, 4, 6, 8, 10)$ can be thought of as a 6-tuple from $\mathbb{N}^6$, or as the function $g: 6 \to \mathbb{N}$ defined by $g(k) = 2k$.

Exercise 5.11: Express each of the following sequences as a k-tuple. What is the value of k?

1. $f: 3 \to \mathbb{Q}$, where $f(n) = \dfrac{n}{3}$ _____ _____

2. $f: 11 \to \mathbb{Z}$, where $f(n) = (-1)^n$ _____ _____

3. $f: 6 \to \mathbb{C}$, where $f(n) = \dfrac{n}{2} + ni$ _____ _____

Example 5.12:

1. There is exactly one finite sequence of length 0, namely the empty sequence, $\emptyset$. The empty sequence can be described using function notation as $f: 0 \to X$, where X is any set.

2. The infinite sequence $(0, -1, 2, -3, 4, -5, \ldots)$ is a function from $\mathbb{N}$ to $\mathbb{Z}$. If we name this function g, then we have that $g: \mathbb{N} \to \mathbb{Z}$ and g is defined by $g(n) = (-1)^n n$. So, the nth term of the sequence g is $(-1)^n n$ and we can represent the sequence as $((-1)^n n)$. This function g is an example of an *integer-valued function* (or *$\mathbb{Z}$-valued function*) because the codomain of g consists of only integers.

3. The infinite sequence $\left(\dfrac{1}{n+1}\right)$ is a function from $\mathbb{N}$ to $\mathbb{Q}$. The nth term of this sequence is $\dfrac{1}{n+1}$. If we name this function h, then we have that $h: \mathbb{N} \to \mathbb{Q}$ and h is defined by $h(n) = \dfrac{1}{n+1}$. This function h is an example of a *rational-valued function* (or *$\mathbb{Q}$-valued function*) because the codomain of h consists of only rational numbers. Since the "outputs" of h take on only positive values, we can "shrink" the codomain of h to $\mathbb{Q}^+$, the set of positive rational numbers. So, we can write $h: \mathbb{N} \to \mathbb{Q}^+$. We can visualize this sequence as follows:

$$\left(1, \frac{1}{2}, \frac{1}{3}, \frac{1}{4}, \frac{1}{5}, \ldots\right)$$

4. The infinite sequence $\left(n^2 + \sqrt{n}i\right)$ is a function from $\mathbb{N}$ to $\mathbb{C}$. The nth term of this sequence is $n^2 + \sqrt{n}i$. If we name this function k, then we have that $k: \mathbb{N} \to \mathbb{C}$ and k is defined by $k(n) = n^2 + \sqrt{n}i$. This function k is an example of a *complex-valued function* (or *$\mathbb{C}$-valued function*) because the codomain of k consists of only complex numbers. We can visualize this sequence as follows: $\left(0, 1 + i, 4 + \sqrt{2}i, 9 + \sqrt{3}i, 16 + 2i, \ldots\right)$

Injections, Surjections, and Bijections

A function $f: A \to B$ is **injective** (or **one-to-one**), written $f: A \hookrightarrow B$, if for all $a, b \in A$, if $a \neq b$, then $f(a) \neq f(b)$. In this case, we call f an **injection**.

Example 5.13: The function $f: \{s, t\} \to \{0, 1\}$ defined by $f = \{(s, 0), (t, 1)\}$ is injective because each "input" goes to a distinct "output." On the other hand, the function $g: \{s, t\} \to \{0, 1\}$ defined by $g = \{(s, 0), (t, 0)\}$ is **not** injective. In the function g, the "inputs" s and t are assigned to the same "output" 0.

Notes: (1) Consider the **conditional** statement $p \to q$ (which we read as "if p, then q" or "p implies q"). The statement $\neg q \to \neg p$ (which we read as "if not p, then not q" or "not p implies not q" is called the **contrapositive** of the original conditional statement. These two statements are **logically equivalent**. This means that either both of these statements are true or both of these statements are false. It follows that we can show that a conditional statement is true by showing that its contrapositive is true instead. See Lesson 7 for more details.

(2) The contrapositive of the statement "If $a \neq b$, then $f(a) \neq f(b)$" is "If $f(a) = f(b)$, then $a = b$." So, we can say that a function $f: A \to B$ is injective if for all $a, b \in A$, if $f(a) = f(b)$, then $a = b$.

A function $f: A \to B$ is **surjective** (or **onto B**), written $f: A \mapsto B$, if for all $b \in B$, there is an $a \in A$ such that $f(a) = b$. In this case, we call f a **surjection**.

Example 5.14: Consider once again the functions $f, g: \{s, t\} \to \{0, 1\}$ from Example 5.13 defined by $f = \{(s, 0), (t, 1)\}$ and $g = \{(s, 0), (t, 0)\}$. The function f is surjective because each element of the codomain $\{0, 1\}$ is in the range of f. On the other hand, the function g is **not** surjective because the element 1 is in the codomain of the function, but not in the range.

A function $f: A \to B$ is **bijective**, written $f: A \cong B$ if f is both injective and surjective. In this case, we call f a **bijection**.

Example 5.15:

1. $f = \{(0, a), (1, a)\}$ from part 1 of Example 5.4 is **not** an injective function because $f(0) = a$, $f(1) = a$, and $0 \neq 1$. If we think of f as $f: \{0, 1\} \to \{a\}$, then f is surjective. However, if we think of f as $f: \{0, 1\} \to \{a, b\}$, then f is **not** surjective. So, surjectivity depends upon the codomain of the function.

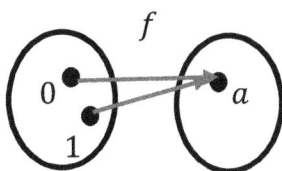

2. $k = \{(a, b) \mid a, b \in \mathbb{R} \wedge b > 0 \wedge a^2 + b^2 = 2\}$ from part 4 of Example 5.4 is **not** an injective function. For example, $(1, 1) \in k$ because $1^2 + 1^2 = 1 + 1 = 2$ and $(-1, 1) \in k$ because $(-1)^2 + 1^2 = 1 + 1 = 2$. Notice how a horizontal line hits the graph twice. If we think of k as a function from $\left(-\sqrt{2}, \sqrt{2}\right)$ to $\mathbb{R}^+$, then k is **not** surjective. For example, $2 \notin \operatorname{ran} k$ because for any $a \in \mathbb{R}$, $a^2 + 2^2 = a^2 + 4 \geq 4$, and so, $a^2 + 2^2$ cannot be equal to 2. However, if instead we consider k as a function with codomain $\left(0, \sqrt{2}\,\right]$, that is $k: \left(-\sqrt{2}, \sqrt{2}\right) \to \left(0, \sqrt{2}\,\right]$, then k **is** surjective. Indeed, if $0 < b \leq \sqrt{2}$, then $0 < b^2 \leq 2$, and so, $a^2 = 2 - b^2 \geq 0$. Therefore, $a = \sqrt{2 - b^2}$ is a real number such that $k(a) = b$.

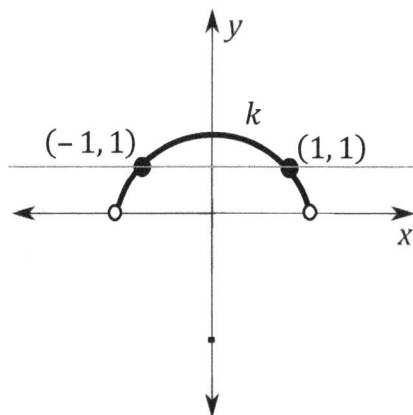

3. Define $g: \mathbb{R} \to \mathbb{R}$ by $g(x) = 7x - 3$. Then g is injective because if $g(a) = g(b)$, we then have $7a - 3 = 7b - 3$. Adding 3 to each side of this equation, we get $7a = 7b$, and then multiplying each side of this last equation by $\frac{1}{7}$, we get $a = b$. Also, g is surjective because if $b \in \mathbb{R}$, then $\frac{b+3}{7} \in \mathbb{R}$ and

$$g\left(\frac{b+3}{7}\right) = 7\left(\frac{b+3}{7}\right) - 3 = (b+3) - 3 = b + (3-3) = b + 0 = b$$

Therefore, g is bijective. See the image to the right for a visual representation of $\mathbb{R}^2$ and the graph of the function g.

Notice that any vertical line will hit the graph of g exactly once because g is a function with domain $\mathbb{R}$. Also, any horizontal line will hit the graph exactly once because g is bijective. Injectivity ensures that each horizontal line hits the graph *at most* once and surjectivity ensures that each horizontal line hits the graph *at least* once.

This function g is an example of a *real-valued function* (or $\mathbb{R}$-*valued function*) because the codomain of g consists of only real numbers.

4. If A is any set, then we define the **identity function** on A, written $i_A: A \to A$, by $i_A(a) = a$ for all $a \in A$. This function is a bijection from A to itself.

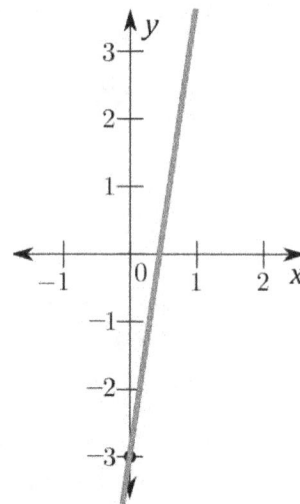

Exercise 5.16: Determine if each of the following functions is bijective, injective only, surjective only, or neither injective nor surjective.

1. $f: \mathbb{Z} \to \mathbb{Z}$, where $f(n) = 2n - 1$ _____

2. $g: \mathbb{R} \to \mathbb{R}$, where $g(r) = 2r - 1$ _____

3. $h: \mathbb{Z} \to \mathbb{N}$, where $h(n) = \begin{cases} n & \text{if } n \geq 0 \\ -n & \text{if } n < 0 \end{cases}$ _____

4. $k: \mathbb{Z} \to \mathbb{Q}$, where $k(n) = \frac{1}{n^2+1}$ _____

Inverse Functions

If a function is bijective, we can take its **inverse** by interchanging the inputs and the outputs.

Example 5.17: The function $f: \{s, t\} \to \{0, 1\}$ defined by $f = \{(s, 0), (t, 1)\}$ is a bijection (see Examples 5.13 and 5.14 above). The inverse of f is the function $g: \{0, 1\} \to \{s, t\}$ defined by $g = \{(0, s), (1, t)\}$. Observe how the domain of f is the range of g and the range of f is the domain of g. Furthermore, we obtain the ordered pairs in g by "reversing" the elements in each ordered pair in f.

Let's now provide a more formal definition of the inverse of a function.

If $f: A \to B$ is bijective, we define $f^{-1}: B \to A$, the **inverse** of f, by $f^{-1} = \{(b, a) \mid (a, b) \in f\}$. In other words, for each $b \in B$, $f^{-1}(b) = $ "the unique $a \in A$ such that $f(a) = b$."

Notes: (1) Let $f: A \to B$ be bijective. Since f is surjective, for each $b \in B$, there is an $a \in A$ such that $f(a) = b$. Since f is injective, there is only one such value of a.

(2) The inverse of a bijective function is also bijective.

(3) The inverse of the inverse of a bijective function is the original function.

Example 5.18:

1. Define $f: \{0, 1\} \to \{a, b\}$ by $f = \{(0, a), (1, b)\}$. Then f is a bijection and $f^{-1}: \{a, b\} \to \{0, 1\}$ is defined by $f^{-1} = \{(a, 0), (b, 1)\}$. Observe that f^{-1} is also a bijection.

2. Let $2\mathbb{N} = \{0, 2, 4, 6, 8, \dots\}$ be the set of even natural numbers and let $2\mathbb{N} + 1 = \{1, 3, 5, 7, 9 \dots\}$ be the set of odd natural numbers. The function $f: 2\mathbb{N} \to 2\mathbb{N} + 1$ defined by $f(n) = n + 1$ is a bijection with inverse $f^{-1}: 2\mathbb{N} + 1 \to 2\mathbb{N}$ defined by $f(n) = n - 1$.

Exercise 5.19: Find the inverse of each of the following functions:

1. $f = \{(a, \square), (b, \Delta), (c, \mho)\}$ _____

2. $g: \mathbb{N} \to 2\mathbb{N}$, where $g(n) = 2n$ _____

3. $h: \mathbb{R} \to \mathbb{R}$, where $h(r) = \frac{2}{3}r + 7$ _____

4. $k: \mathbb{C} \to \mathbb{C}$, where $k(z) = iz$ _____

If X and Y are sets, we define ^XY to be the set of functions from X to Y. Symbolically, we have

$$^XY = \{f \mid f: X \to Y\}.$$

Example 5.20: If $A = \{a, b\}$ and $2 = \{0, 1\}$, then $^A2$ has 4 elements (each element is a function from A to 2). The elements of $^A2$ are as follows: $f_1 = \{(a, 0), (b, 0)\}$, $f_2 = \{(a, 0), (b, 1)\}$, $f_3 = \{(a, 1), (b, 0)\}$, and $f_4 = \{(a, 1), (b, 1)\}$. Here is a visual representation of these four functions.

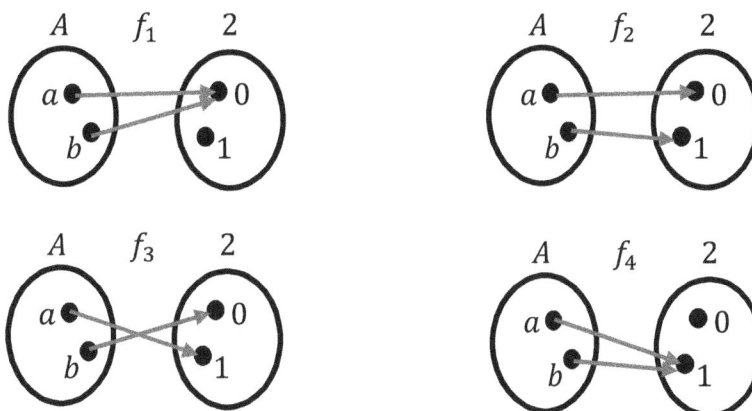

Recall from Lesson 1 that the power set of A, written $\mathcal{P}(A)$, consists of all the subsets of A. So, in this example, $\mathcal{P}(A) = \{\emptyset, \{a\}, \{b\}, \{a, b\}\}$.

We can define a bijection F from $^A2$ to $\mathcal{P}(A)$ as follows:

$F: {}^A2 \to \mathcal{P}(A)$ is defined by $F(f) = \{x \in A \mid f(x) = 1\}$. In other words, we associate each function in ${}^A2$ with the set of elements whose image is 1.

So, $F(f_1) = \emptyset$, $F(f_2) = \{b\}$, $F(f_3) = \{a\}$, and $F(f_4) = \{a, b\}$.

Since $\mathcal{P}(A) = \{\emptyset, \{a\}, \{b\}, \{a, b\}\}$, we see that F is a bijection from ${}^A2$ to $\mathcal{P}(A)$

The inverse of F is the function $F^{-1}: \mathcal{P}(A) \to {}^A B$ defined by $F^{-1}(C)(x) = \begin{cases} 0 & \text{if } x \notin C. \\ 1 & \text{if } x \in C. \end{cases}$ In other words, we associate each subset C of A with the function that sends elements of C to 1 and the rest of the elements to 0.

So, we see that $F^{-1}(\emptyset) = f_1$, $F^{-1}(\{b\}) = f_2$, $F^{-1}(\{a\}) = f_3$, and $F^{-1}(\{a, b\}) = f_4$.

Exercise 5.21: Let A be a nonempty set, let $2 = \{0, 1\}$, and define the function $F: {}^A2 \to \mathcal{P}(A)$ by $F(f) = \{x \in A \mid f(x) = 1\}$.

1. Explain why F is injective.

_____

_____

_____

2. Explain why F is surjective.

_____

_____

_____

3. Explain why F is bijective.

_____

_____

_____

4. What is the inverse of F?   _____

Composite Functions

Given a function $f: A \to B$, we "apply" the function f to an "input" $a \in A$ to get an "output" $b \in B$. If we have another function $g: B \to C$. We can now think of b as an input for g. "Applying" the function g to this input b gives us some output $c \in C$. This procedure of "applying" one function followed by another function is called "composition."

Formally, given functions $f: A \to B$ and $g: B \to C$, the **composite** (or **composition**) of f and g, written $g \circ f: A \to C$, is defined by $(g \circ f)(a) = g(f(a))$ for all $a \in A$. Symbolically, we have

$$g \circ f = \{(a, c) \in A \times C \mid \text{There is a } b \in B \text{ such that } (a, b) \in f \text{ and } (b, c) \in g\}.$$

We can visualize the composition of two functions f and g as follows.

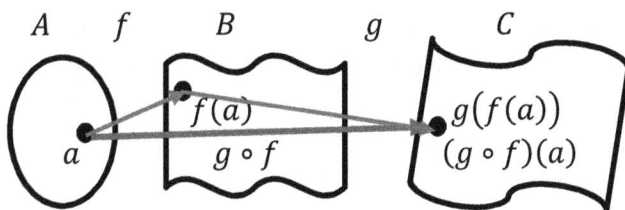

In the picture above, sets A, B, and C are drawn as different shapes simply to emphasize that they can all be different sets. Starting with an arbitrary element $a \in A$, we have an arrow showing a being mapped by f to $f(a) \in B$ and another arrow showing $f(a)$ being mapped by g to $g(f(a)) \in C$. There is also an arrow going directly from $a \in A$ to $(g \circ f)(a) = g(f(a))$ in C. However, note that the only way we know how to get from a to $(g \circ f)(a)$ is to first travel from a to $f(a)$, and then to travel from $f(a)$ to $g(f(a))$.

Example 5.22: Define $f: \mathbb{R} \to \mathbb{R}$ by $f(x) = x + 7$ and $g: \mathbb{R} \to \mathbb{R}$ by $g(x) = (x - 5)^2$ Then $g \circ f: \mathbb{R} \to \mathbb{R}$ is defined by $(g \circ f)(x) = g(f(x)) = g(x + 7) = ((x + 7) - 5)^2 = (x + 2)^2$.

Exercise 5.23: Define $f: \mathbb{Z} \to \mathbb{Q}$ by $f(n) = \frac{n}{2}$ and define $g: \mathbb{Q} \to \{0, 1\}$ by $g(n) = \begin{cases} 0 & \text{if } n \in \mathbb{Z}. \\ 1 & \text{if } n \in \mathbb{Q} \setminus \mathbb{Z}. \end{cases}$

1. Assuming that n is even, evaluate $(g \circ f)(n)$. _____

2. Assuming that n is odd, evaluate $(g \circ f)(n)$. _____

3. Provide an explicit definition of the function $g \circ f$. _____

The following basic facts about function compositions are useful.

Composition Fact 1: The composition of injective functions is injective.

Composition Fact 2: The composition of surjective functions is surjective.

Composition Fact 3: The composition of bijective functions is bijective.

Composition Fact 4: If $f: A \cong B$, then $f^{-1} \circ f = i_A$ and $f \circ f^{-1} = i_B$ (recall that $i_A: A \to A$ is the identity function on A, and similarly, $i_B: B \to B$ is the identity function on B).

You will be asked to verify these facts in Problems 76 – 80 below.

The Real Numbers

Note: This section will not be of any importance in this book. I am including it to provide a concrete definition of the real numbers for those readers that are interested. If you continue to think of the real numbers informally, as we have been doing up until this point, you will not be at a disadvantage when reading any other part of this book.

There are several equivalent ways to define the set of real numbers. We will define this set here as equivalence classes of **Cauchy sequences**. For our purposes, an informal description of a Cauchy sequence will suffice. Informally, a Cauchy sequence is a *rational-valued sequence* whose values get "closer and closer to each other" as we go further out into the sequence.

Note: We can recognize a Cauchy sequence as one that <u>seems</u> to be "converging" to a fixed value. However, this "value" does not need to be a rational number.

Example 5.24:

1. The sequence $(x_n) = \left(\frac{1}{n+1}\right) = \left(1, \frac{1}{2}, \frac{1}{3}, \frac{1}{4}, \frac{1}{5}, \dots\right)$ is a Cauchy sequence, as it seems to be converging to 0.

2. The sequence $(y_n) = (n) = (0, 1, 2, 3, 4, 5, \dots)$ is not a Cauchy sequence, as it is **not** converging to a fixed real number.

3. The sequence $(s_n) = (1, 1.4, 1.41, 1.414, 1.4142, 1.41421, 1.414213, \dots)$ consists of finite approximations to the "real number" $\sqrt{2}$ (estimate $\sqrt{2}$ with your calculator so that you can see this for yourself). This is an example of a Cauchy sequence that seems to be converging, but is not actually converging to a rational number.

Exercise 5.25: Determine if each of the following is a Cauchy sequence. If so, determine to what number it seems to be converging. Is the number that it is converging to a rational number?

1. $(a_n) = \left(1 - \frac{1}{n+1}\right)$ _____

2. $(b_n) = (0, 1, 0, 2, 0, 3, 0, 4, \dots)$ _____

3. $(c_n) = (5^n)$ _____

4. $(d_n) = \left(\frac{1}{7^{n+1}}\right)$ _____

5. $(e_n) = (3, 3.1, 3.14, 3.141, 3.1415, 3.14159, \dots)$ _____

Next, we would like to identify Cauchy sequences that seem to be converging to the same value. For example, we will identify the Cauchy sequences $(x_n) = \left(\frac{1}{n+1}\right)$ and $(y_n) = (0)$.

An equivalent way of saying that (x_n) and (y_n) converge to the same value is to say that $(x_n - y_n)$ converges to 0.

Let $A = \{(x_n) \mid (x_n) \text{ is a Cauchy sequence of rational numbers}\}$. We define an equivalence relation R on A by:

$$(x_n)R(y_n) \text{ if and only if } (x_n - y_n) \text{ converges to } 0.$$

We define the set of real numbers as follows:

$$\mathbb{R} = \{[(x_n)] \mid (x_n) \text{ is a Cauchy sequence of rational numbers}\}.$$

We identify the real number $[(q)]$ with the rational number q. In this way, we have $\mathbb{Q} \subseteq \mathbb{R}$.

Problem Set 5

Full solutions to these problems are available for free download here:

www.SATPrepGet800.com/STKZ3D

LEVEL 1

Determine if each of the following relations is a function. State the domain and range of each such function.

1. $\{(0, 0), (1, 0)\}$

2. $\{(0, 0), (0, 1)\}$

3. $\{(a, a), (b, b), (c, c)\}$

4. $\{(a, a), (a, b), (a, c)\}$

5. $\{(a, a), (b, a), (c, a)\}$

6. $\{(a, b), (b, b), (c, d), (e, a)\}$

7. $\{(a, a), (a, b), (b, a)\}$

8. $\{(0, 0), (1, 1), (2, 2), (3, 3), (4, 4)\}$

9. $\{(0, a), (a, 0), (b, 1), (1, c), (c, 0)\}$

10. $\{(0, a), (a, 0), (0, 1), (b, 2)\}$

Determine if each of the following functions is injective.

11. $\{(s, 0), (t, 0)\}$

12. $\{(s, 0), (t, 1)\}$

13. $\{(a, a), (b, b), (c, c)\}$

14. $\{(a, a), (b, a), (c, b)\}$

15. $\{(s, s), (t, s), (u, s)\}$

16. $\{(a, b), (b, a), (c, d), (d, c)\}$

Determine if each of the following relations is a function. State the domain and range of each such function.

17. $\{(a, b) \in \mathbb{R}^2 \mid b \leq 0 \wedge a^2 + b^2 = 9\}$

18. $\{(a, b) \in \mathbb{R}^2 \mid a < 0 \wedge a^2 + b^2 = 9\}$

19. $\left\{\left(\frac{a}{b}, c\right) \in \mathbb{Q} \times \mathbb{Z} \mid c = a - b\right\}$

20. $\{(a + bi, c) \in \mathbb{C} \times \mathbb{R} \mid c = a\}$

21. $\{(c, a + bi) \in \mathbb{R} \times \mathbb{C} \mid c = \max\{a, b\}\}$, where $\max\{a, b\}$ is the larger of a or b.

Determine if each of the following functions is injective, surjective, both, or neither.

22. $\{(a, b) \in \mathbb{R}^2 \mid b > 0 \wedge a^2 + b^2 = 1\}$

23. $\{(a, b) \in \mathbb{R} \times (0, 1] \mid b > 0 \wedge a^2 + b^2 = 1\}$

24. $\{(a + bi, c) \in \mathbb{C} \times \mathbb{R} \mid c = a\}$

25. $\{(n, m) \in \mathbb{N}^2 \mid m = 5n\}$

26. $\{(n, z) \in \mathbb{Z} \times \mathbb{C} \mid z = n + ni\}$

If $f : X \to Y$ and $A \subseteq X$, then the **image of A under f** is the set $f[A] = \{f(x) \mid x \in A\}$. Let $f : \{a, b, c, d\} \to \{0, 1, 2\}$ be defined by $f = \{(a, 0), (b, 0), (c, 1), (d, 2)\}$. Compute $f[A]$ for each of the following sets A:

27. $A = \{a\}$

28. $A = \{a, b\}$

29. $A = \{a, b, c\}$

30. $A = \{a, c\}$

31. $A = \{b, c, d\}$

If $B \subseteq Y$, then the **inverse image of B under f** is the set $f^{-1}[B] = \{x \in X \mid f(x) \in B\}$. Let $f:\{a, b, c, d\} \to \{0, 1, 2\}$ be defined by $f = \{(a, 0), (b, 0), (c, 1), (d, 2)\}$. Compute $f^{-1}[B]$ for each of the following sets B:

32. $B = \{0\}$

33. $B = \{1\}$

34. $B = \{0, 1\}$

35. $B = \{0, 2\}$

36. $B = \{0, 1, 2\}$

LEVEL 3

If $f: X \to Y$ and $A \subseteq X$, then the **image of A under f** is the set $f[A] = \{f(x) \mid x \in A\}$. Let $f: \mathbb{R} \to \mathbb{R}$ be defined by $f(x) = x^4$. Compute $f[A]$ for each of the following sets A:

37. $A = \mathbb{R}$

38. $A = \{-2, 0, 3\}$

39. $A = (-3, 2]$

40. $A = (4, \infty)$

41. $A = (-\infty, 1]$

If $B \subseteq Y$, then the **inverse image of B under f** is the set $f^{-1}[B] = \{x \in X \mid f(x) \in B\}$. Let $f: \mathbb{R} \to \mathbb{R}$ be defined by $f(x) = x^4$. Compute $f^{-1}[B]$ for each of the following:

42. $B = \{16\}$

43. $B = \{-1\}$

44. $B = [0, \infty)$

45. $B = (-\infty, 0)$

46. $B = (0, \infty)$

For each of the following sets X and Y, compute ^XY:

47. $X = \emptyset, Y = \emptyset$

48. $X = \{a\}, Y = \{b\}$

49. $X = \{a, b\}, Y = \{c\}$

50. $X = \{a, b\}, Y = \{a\}$

51. $X = \{a\}, Y = \{0, 1\}$

52. $X = \{a, b\}, Y = \{a, b, c\}$

53. $X = \{a, b, c\}, Y = \{0, 1\}$

54. $X = \{a, b, c\}, Y = \{0, 1, 2\}$

LEVEL 4

For $f, g \in {^\mathbb{R}\mathbb{R}}$, define $f \preccurlyeq g$ if and only if for all $x \in \mathbb{R}$, $f(x) \le g(x)$.

55. Is $\preccurlyeq$ reflexive on $^\mathbb{R}\mathbb{R}$?

56. Is $\preccurlyeq$ symmetric on $^\mathbb{R}\mathbb{R}$?

57. Is $\preccurlyeq$ transitive on $^\mathbb{R}\mathbb{R}$?

58. Is $\preccurlyeq$ antireflexive on $^\mathbb{R}\mathbb{R}$?

59. Is $\preccurlyeq$ antisymmetric on $^\mathbb{R}\mathbb{R}$?

60. Is $\preccurlyeq$ trichotomous on $^\mathbb{R}\mathbb{R}$?

61. Does $\preccurlyeq$ satisfy the comparability condition on $^\mathbb{R}\mathbb{R}$?

62. Is $\preccurlyeq$ a partial ordering on $^\mathbb{R}\mathbb{R}$?

63. Is $\preccurlyeq$ a linear ordering on $^\mathbb{R}\mathbb{R}$?

For $f, g \in {}^{\mathbb{R}}\mathbb{R}$, define $f \preccurlyeq^* g$ if and only if there is an $x \in \mathbb{R}$ such that $f(x) \leq g(x)$.

64. Is $\preccurlyeq^*$ reflexive on ${}^{\mathbb{R}}\mathbb{R}$?

65. Is $\preccurlyeq^*$ symmetric on ${}^{\mathbb{R}}\mathbb{R}$?

66. Is $\preccurlyeq^*$ transitive on ${}^{\mathbb{R}}\mathbb{R}$?

67. Is $\preccurlyeq^*$ antireflexive on ${}^{\mathbb{R}}\mathbb{R}$?

68. Is $\preccurlyeq^*$ antisymmetric on ${}^{\mathbb{R}}\mathbb{R}$?

69. Is $\preccurlyeq^*$ trichotomous on ${}^{\mathbb{R}}\mathbb{R}$?

70. Does $\preccurlyeq^*$ satisfy the comparability condition on ${}^{\mathbb{R}}\mathbb{R}$?

71. Is $\preccurlyeq^*$ a partial ordering on ${}^{\mathbb{R}}\mathbb{R}$?

72. Is $\preccurlyeq^*$ a linear ordering on ${}^{\mathbb{R}}\mathbb{R}$?

LEVEL 5

Determine if each of the following sequences is a Cauchy sequence. Are any of the Cauchy sequences equivalent?

73. $(x_n) = \left(1 + \frac{1}{n+1}\right)$

74. $(y_n) = (2^n)$

75. $(z_n) = \left(1 - \frac{1}{2n+1}\right)$

Let $f: A \to B$ and $g: B \to C$. Show that each of the following is true.

76. If f is injective and g is injective, then $g \circ f$ is injective.

77. If f is surjective and g is surjective, then $g \circ f$ is surjective.

78. If f is bijective and g is bijective, then $g \circ f$ is bijective.

Let $f: A \cong B$. Show that each of the following is true.

79. If f is bijective, then $f^{-1} \circ f = i_A$.

80. If f is bijective, then $f \circ f^{-1} = i_B$.

CHALLENGE PROBLEMS

81. For $f, g \in {}^{\mathbb{N}}\mathbb{N}$, define $f <^* g$ if and only if there is $n \in \mathbb{N}$ such that for all $m > n, f(m) < g(m)$. Determine if $<^*$ a partial ordering on ${}^{\mathbb{R}}\mathbb{R}$. If so, is $<^*$ a linear ordering on ${}^{\mathbb{R}}\mathbb{R}$?

82. Recall that we formally define the integers to be the set $\mathbb{Z} = \{[(a,b)] \mid (a,b) \in \mathbb{N} \times \mathbb{N}\}$, where $[(a,b)]$ is the equivalence class of $R = \{((a,b),(c,d)) \in (\mathbb{N} \times \mathbb{N})^2 \mid a + d = b + c\}$. Define functions $f: \mathbb{Z} \times \mathbb{Z} \to \mathbb{Z}$ and $g: \mathbb{Z} \times \mathbb{Z} \to \mathbb{Z}$ by $f([(a,b)], [(c,d)]) = [(a+c, b+d)]$ and $g([(a,b)], [(c,d)]) = [(ac + bd, ad + bc)]$. Show that these functions are well-defined.

LESSON 6
EQUINUMEROSITY

Equinumerous Sets

We say that two sets A and B are **equinumerous**, written $A \sim B$, if there is a bijection $f: A \cong B$. In this case, we may also say that A and B have the same **cardinality**, and we can write $|A| = |B|$.

Example 6.1:

1. Let $A = \{\text{anteater}, \text{elephant}, \text{giraffe}\}$ and $B = \{\text{apple}, \text{banana}, \text{orange}\}$. Then $A \sim B$. We can define a bijection $f: A \cong B$ by $f(\text{anteater}) = \text{apple}$, $f(\text{elephant}) = \text{banana}$, and $f(\text{giraffe}) = \text{orange}$. This is not the only bijection from A to B (there are 6 distinct bijections from A to B), but we need only find one (or verify that one exists) to show that the sets are equinumerous.

2. Two finite sets are equinumerous if and only if they have the same number of elements. Furthermore, a finite set can never be equinumerous with an infinite set.

3. Let $\mathbb{N} = \{0, 1, 2, 3, 4 \dots\}$ be the set of natural numbers and let $2\mathbb{N} = \{0, 2, 4, 6, 8 \dots\}$ be the set of even natural numbers. Then $\mathbb{N} \sim 2\mathbb{N}$. We can actually see a bijection between these two sets just by looking at the sets themselves.

$$0 \quad 1 \quad 2 \quad 3 \quad 4 \quad 5 \quad 6 \dots$$
$$0 \quad 2 \quad 4 \quad 6 \quad 8 \quad 10 \quad 12 \dots$$

   The function $f: \mathbb{N} \to 2\mathbb{N}$ defined by $f(n) = 2n$ is an explicit bijection.

4. Is $\mathbb{N} \sim \mathbb{Z}$? We can see a bijection between these two sets by listing the elements of the sets in a particular order. We list the elements of $\mathbb{N}$ in their natural order (which we will usually do). However, we will need to be a bit more clever in how we list the elements of $\mathbb{Z}$.

$$0 \quad 1 \quad 2 \quad 3 \quad 4 \quad 5 \quad 6 \dots$$
$$0 \quad -1 \quad 1 \quad -2 \quad 2 \quad -3 \quad 3 \dots$$

Observe how we listed the elements of $\mathbb{Z}$ by placing 0 first and then alternating between natural numbers and their negatives.

Many students get confused here because they are under the misconception that the integers should be written "in order." However, when checking to see if two sets are equinumerous, we **do not** include any other structure. In other words, we are just trying to "pair up" elements—it does not matter how we do so.

An explicit bijection $f: \mathbb{N} \cong \mathbb{Z}$ can be defined by $f(n) = \begin{cases} \dfrac{n}{2} & \text{if } n \text{ is even.} \\ -\dfrac{n+1}{2} & \text{if } n \text{ is odd.} \end{cases}$

You will be asked to verify that f is a bijection in Problems 29 – 31 below.

Exercise 6.2: For each of the following, determine if A and B are equinumerous. If so, provide a bijection from one set to the other. If not, explain why.

1. $A = \{a, b, c, d\}$, $B = \{x, y, z, w\}$

   _____

   _____

2. $A = \{0, 1, 2, 3, \dots, 99, 100\}$, $B = \{1, 2, 3, \dots, 101, 102\}$

   _____

   _____

3. $A = \{5, 6, 7, \dots, 8020, 8021\}$, $B = \mathbb{N}$

   _____

   _____

4. $A = \mathbb{Z}$, $B = 2\mathbb{Z}$

   _____

   _____

5. $A = {}^{\mathbb{N}}2$, $B = \mathcal{P}(\mathbb{N})$

   _____

   _____

6. $A = [0, 1]$, $B = [0, 3]$

   _____

   _____

You may have noticed that we used the expression "A and B are equinumerous," rather than "A is equinumerous with B." Although this latter expression is equally correct, we can use the first one because equinumerosity is an equivalence relation.

Exercise 6.3: Show that equinumerosity is an equivalence relation by verifying each of the following:

1. $\sim$ is reflexive.

   _____

   _____

2. $\sim$ is symmetric.

   _____

   _____

3. $\sim$ is transitive.

_____

_____

Countable Sets

We say that a set is **countable** if it is equinumerous with a subset of $\mathbb{N}$. It's easy to visualize a countable set because a bijection from a subset of $\mathbb{N}$ to a set A generates a list. For example, the set $2\mathbb{N}$ can be listed as $0, 2, 4, 6, \ldots$ and the set $\mathbb{Z}$ can be listed as $0, -1, 1, -2, 2, \ldots$ (see Example 6.1 above).

There are two kinds of countable sets: finite sets and **denumerable** sets. We say that a set is denumerable if it is countably infinite.

Example 6.4:

1. The set $A = \{\text{anteater}, \text{elephant}, \text{giraffe}\}$ is countable because there is a bijection from the set $3 = \{0, 1, 2\}$ to A given by $\{(0, \text{anteater}), (1, \text{elephant}), (2, \text{giraffe})\}$. This is an example of a countable set that is finite.

2. The set $\mathbb{N}$ is denumerable, as the identity function $i_{\mathbb{N}}:\mathbb{N} \to \mathbb{N}$ is a bijection.

3. We saw in parts 3 and 4 of Example 6.1 that $2\mathbb{N}$ and $\mathbb{Z}$ are denumerable sets.

4. Let's show that the set of positive rational numbers, $\mathbb{Q}^+$, is denumerable. This is a bit trickier than previous examples. We begin by arranging the elements of $\mathbb{Q}^+$ in the following array:

$$\frac{1}{1} \quad \frac{2}{1} \quad \frac{3}{1} \quad \frac{4}{1} \quad \frac{5}{1} \quad \ldots$$

$$\frac{1}{2} \quad \frac{2}{2} \quad \frac{3}{2} \quad \frac{4}{2} \quad \frac{5}{2} \quad \ldots$$

$$\frac{1}{3} \quad \frac{2}{3} \quad \frac{3}{3} \quad \frac{4}{3} \quad \frac{5}{3} \quad \ldots$$

$$\frac{1}{4} \quad \frac{2}{4} \quad \frac{3}{4} \quad \frac{4}{4} \quad \frac{5}{4} \quad \ldots$$

$$\ldots \quad \ldots \quad \ldots \quad \ldots$$

The ellipses to the right indicate that each row consists of infinitely many rational numbers. Similarly, the ellipses at the bottom indicate that each column consists of infinitely many rational numbers.

Observe how the first row consists of all positive rational numbers with denominator 1, the second row consists of all positive rational numbers with denominator 2, and so on. In this way, we see that every positive rational number appears somewhere in the array.

The following image will show us how to turn this array into a single list:

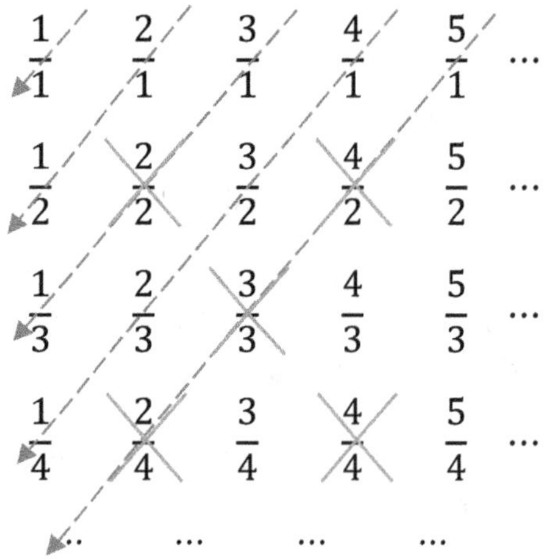

We now form a list of the positive rational numbers by beginning in the upper left corner, following each arrow, moving to the right each time we complete the path of an arrow, and ignoring the elements that are x'd out.

The first arrow indicates that we should write the rational number $\frac{1}{1}$ first. As usual, we will abbreviate this as 1.

We now move to the right to the next arrow. That arrow indicates that we should write the rational number $\frac{2}{1} = 2$, followed by the rational number $\frac{1}{2}$. So far, our list looks as follows:

$$1, 2, \frac{1}{2}$$

The next arrow to the right indicates that we should write the rational number $\frac{3}{1} = 3$, followed by the rational number $\frac{1}{3}$. Notice that we skipped $\frac{2}{2} = 1$ because we already wrote that one down. Remember that each rational number has infinitely many representations, and so, we will often need to skip rational numbers that we have already accounted for. So far, our list looks as follows:

$$1, 2, \frac{1}{2}, 3, \frac{1}{3}$$

Continuing in this fashion, we generate the following list of the positive rational numbers:

$$1, 2, \frac{1}{2}, 3, \frac{1}{3}, 4, \frac{3}{2}, \frac{2}{3}, \frac{1}{4}, 5, \frac{1}{5}, \dots$$

Exercise 6.5: Show that each of the following sets is denumerable by generating a list.

1. $2\mathbb{N} + 1 = \{1, 3, 5, 7, \dots\}$

_____

2. $2\mathbb{Z} + 1 = \{\dots, -7, -5, -3, -1, 1, 3, 5, 7, \dots\}$

_____

3. $\mathbb{Q} = \left\{ \frac{a}{b} \mid a, b \in \mathbb{Z} \text{ and } b \neq 0 \right\}$

Uncountable Sets

At this point, you may be asking yourself if all infinite sets are denumerable. If this were the case, then we would simply have finite sets and infinite sets, and that would be the end of it. However, there are in fact infinite sets that are **not** denumerable. An infinite set that is not denumerable is **uncountable**.

Recall from Lesson 1 that the power set of A, written $\mathcal{P}(A)$, is the set of all subsets of A.

> **Equinumerosity Fact 1:** If A is any set, then A and $\mathcal{P}(A)$ are **not** equinumerous (this result is known as **Cantor's Theorem**).

> **Equinumerosity Fact 2:** If A is any set, then A is equinumerous with a subset of $\mathcal{P}(A)$.

The two facts given above suggest that for any set A, the power set of A, $\mathcal{P}(A)$, is a set that is "larger in size" than the set A.

Example 6.6:

1. Let $A = \{0, 1\}$. Then $\mathcal{P}(A) = \{\emptyset, \{0\}, \{1\}, \{0, 1\}\}$. Observe that $|A| = 2$, whereas $|\mathcal{P}(A)| = 4$. So, A and $\mathcal{P}(A)$ are **not** equinumerous.

2. Recall from part 4 of Example 1.29 that if $|A| = n$, then $|\mathcal{P}(A)| = 2^n$. Since $2^n > n$ for all natural numbers n, we see that for any finite set A, A and $\mathcal{P}(A)$ are not equinumerous.

3. By Equinumerosity Fact 1, $\mathbb{N}$ is not equinumerous with $\mathcal{P}(\mathbb{N})$. By Equinumerosity Fact 2, $\mathbb{N}$ is equinumerous with a subset of $\mathcal{P}(\mathbb{N})$ (in fact there are many such subsets). Here is an example of such a subset:

$$\{\{0\}, \{1\}, \{2\}, \{3\}, \{4\}, \dots\}$$

Here is a visualization of a bijection from $\mathbb{N}$ to this subset of $\mathcal{P}(\mathbb{N})$.

$$\begin{array}{ccccc} 0 & 1 & 2 & 3 & 4 \dots \\ \{0\} & \{1\} & \{2\} & \{3\} & \{4\} \dots \end{array}$$

Observe that we are matching up each natural number with a subset of natural numbers (a very simple subset consisting of just one natural number) in a way so that different natural numbers get matched with different subsets. In other words, we defined an injective function from $\mathbb{N}$ to $\mathcal{P}(\mathbb{N})$. It seems like there are lots of subsets of $\mathbb{N}$ that didn't get mapped to (for example, all infinite subsets of $\mathbb{N}$). So, in a sense, it seems that $\mathbb{N}$ might be a "smaller" set than $\mathcal{P}(\mathbb{N})$.

Exercise 6.7: Let A be any set.

1. Describe a subset B of $\mathcal{P}(A)$ such that A and B are equinumerous. _____

2. Define a bijection $f: A \to B$.   _____

We use the notation $A \preccurlyeq B$ if there is an injective function from A to B.

$$A \preccurlyeq B \text{ if and only if } \exists f\,(f: A \hookrightarrow B)$$

Notes: (1) The symbol $\exists$ is called an **existential quantifier**, and it is pronounced "There exists" or "There is." The expression $\exists f\,(f: A \hookrightarrow B)$ can be translated into English as "There exists an f such that f is an injective function from A to B." See Lesson 7 for more information on quantifiers.

(2) Another way to express $A \preccurlyeq B$ is to say that A is equinumerous with a subset of B.

We write $A \prec B$ if $A \preccurlyeq B$ and $A \nsim B$.

Example 6.8:

1. If A is any set, then $A \prec \mathcal{P}(A)$. For example, $\mathbb{N} \prec \mathcal{P}(\mathbb{N})$.

2. If we let $A = \mathcal{P}(\mathbb{N})$, we can apply part 1 of this example to this set A to see that $\mathcal{P}(\mathbb{N}) \prec \mathcal{P}\big(\mathcal{P}(\mathbb{N})\big)$. Continuing in this fashion, we get a sequence of increasingly larger sets.

$$\mathbb{N} \prec \mathcal{P}(\mathbb{N}) \prec \mathcal{P}\big(\mathcal{P}(\mathbb{N})\big) \prec \mathcal{P}\left(\mathcal{P}\big(\mathcal{P}(\mathbb{N})\big)\right) \prec \cdots$$

Exercise 6.9: Use one of the symbols $\prec$ or $\sim$ to describe the relationship between A and B:

1. $A = \{a, b\}, B = \mathcal{P}(\{a, b\})$ _____

2. $A = \{a, b\}, B = \mathcal{P}(\{a\})$ _____

3. $A = \mathcal{P}(\{0, 1, 2, \ldots, 9001, 9002\}), B = \mathbb{Z}$ _____

4. $A = \mathbb{Z}, B = \mathcal{P}(\mathbb{Z})$ _____

5. $A = \mathbb{Z}, B = \mathcal{P}(\mathbb{N})$ _____

6. $A = \mathcal{P}(\mathbb{Q}), B = \mathcal{P}\big(\mathcal{P}(2\mathbb{Z})\big)$ _____

Cantor-Schroeder-Bernstein

If A and B are arbitrary sets, in general it can be difficult to determine if A and B are equinumerous by producing a bijection. Luckily, the next fact provides an easier way.

> **Equinumerosity Fact 3:** If A and B are sets such that $A \preccurlyeq B$ and $B \preccurlyeq A$, then $A \sim B$ (this result is known as the Cantor-Schroeder-Bernstein Theorem).

Note: At first glance, many students think that the Cantor-Schroeder-Bernstein Theorem (Equinumerosity Fact 3 above) is obvious. This is not true. The theorem says that if there is an injective function from A to B and another injective function from B to A, then there is a bijective function from A to B. This is a deep result, which is far from obvious. Constructing a bijection from two arbitrary injections is not an easy thing to do. The interested reader may want to take a few minutes to try to do it, if for no other reason than to convince themselves that it is difficult.

Example 6.10: Let's use the Cantor-Shroeder-Bernstein Theorem (Equinumerosity Fact 3) to show that the open interval of real numbers $(0, 1)$ is equinumerous with the closed interval of real numbers $[0, 1]$.

Since $(0, 1) \subseteq [0, 1]$, there is an obvious injective function $f: (0, 1) \to [0, 1]$ (just send each element to itself). This shows that $(0, 1) \preccurlyeq [0, 1]$

The harder direction is finding an injective function g from $[0, 1]$ into $(0, 1)$. We will do this by drawing a line segment with endpoints $\left(0, \frac{1}{4}\right)$ and $\left(1, \frac{3}{4}\right)$. This will give us a bijection from $[0, 1]$ to $\left[\frac{1}{4}, \frac{3}{4}\right]$. We can visualize this bijection using the graph to the right. We will write an equation for this line segment in the slope-intercept form $y = mx + b$. Here m is the slope of the line and b is the y-intercept of the line. We can use the graph to see that $b = \frac{1}{4}$ and

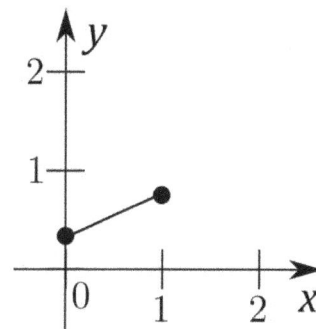

$m = \frac{\text{rise}}{\text{run}} = \frac{\frac{3}{4} - \frac{1}{4}}{1 - 0} = \frac{2}{4} = \frac{1}{2}$. So, we define $g: [0, 1] \to (0, 1)$ by $g(x) = \frac{1}{2}x + \frac{1}{4}$.

This function g is an injection, as can be seen by the graph (it passes the horizontal line test). This shows that $[0, 1] \preccurlyeq (0, 1)$.

Since $(0, 1) \preccurlyeq [0, 1]$ and $[0, 1] \preccurlyeq (0, 1)$, by the Cantor-Schroeder-Bernstein Theorem, $(0, 1) \sim [0, 1]$.

Notes: (1) If $A \subseteq B$, then the function $f: A \to B$ defined by $f(a) = a$ for all $a \in A$ is always injective. It is called the **inclusion map**.

(2) It is unfortunate that the same notation is used for points (ordered pairs of real numbers) and open intervals. Normally this isn't an issue, but in Example 6.10 above both usages of this notation appear. Take another look at Example 6.10 and make sure you can see when the notation (a, b) is being used for a point and when it is being used for an open interval.

(3) We could have used any closed interval $[a, b]$ with $0 < a < b < 1$ in place of $\left[\frac{1}{4}, \frac{3}{4}\right]$.

Exercise 6.11: Let A be the open interval $(1, 5)$ and let B be the closed interval $[0, 6]$.

1. Show that $A \preccurlyeq B$. _____

2. Show that $B \preccurlyeq A$. _____

_____

3. Explain why $A \sim B$. _____

Problem Set 6

Full solutions to these problems are available for free download here:

www.SATPrepGet800.com/STKZ3D

LEVEL 1

For each of the following, determine if A and B are equinumerous. If so, provide a bijection from one set to the other. If not, explain why.

1. $A = \emptyset, B = \{\emptyset\}$

2. $A = \mathcal{P}(\emptyset), B = \{\emptyset\}$

3. $A = \{0, 1, 2\}, B = \{\Delta, \square, \perp\}$

4. $A = \{n \in \mathbb{N} \mid 5 \leq n \leq 250\}, B = \{6, 7, 8, \ldots, 250, 251\}$

5. $A = \{2n \mid n \in \mathbb{Z}\}, B = \{2n \in \mathbb{Z} \mid -100 < n < 9000\}$

6. $A = {}^{\mathbb{Z}}2, B = \mathcal{P}(\mathbb{Z})$

7. $A = {}^{\mathbb{Q}}2, B = \mathcal{P}(\mathbb{Z})$ (you may use Problem 63 below)

8. $A = (0, 1), B = (0, 2)$

State whether each of the following sets is finite, denumerable, or uncountable.

9. $\{3, 4, 5, \ldots\}$

10. $\{3, 4, 5, \ldots, 888, 889\}$

11. $\mathcal{P}(\{3, 4, 5, \ldots\})$

12. $\mathcal{P}(\{3, 4, 5, \ldots, 888, 889\})$

13. $\mathcal{P}\big(\mathcal{P}(\{3, 4, 5, \ldots, 888, 889\})\big)$

14. $\bigcup\{\{n\} \mid n \in \mathbb{Z}\}$

15. $\mathbb{Z} \setminus \mathbb{N}$

16. $\mathcal{P}(\mathbb{Q} \setminus \mathbb{Z})$

LEVEL 2

Show that each of the following sets is denumerable by generating a list.

17. $3\mathbb{N}$

18. $3\mathbb{Z}$

19. $\{n \in \mathbb{N} \mid n$ is not divisible by $5\}$

20. $\{n \in \mathbb{Z} \mid n$ is not divisible by $5\}$

21. $\mathbb{Q}^-$

22. $\mathbb{N} \times \mathbb{N}$

23. $\mathbb{Z} \times \mathbb{Z}$

Use one of the symbols $\prec$ or $\sim$ to describe the relationship between A and B.

24. $A = \{0, 1, 2\}, B = \{1, 2, 3, 4\}$

25. $A = \{0, 1, 2\}, B = \mathcal{P}(\{0, 1, 2\})$

26. $A = \mathbb{Q}, B = \mathcal{P}(\mathbb{Q})$

27. $A = \mathcal{P}(\mathbb{N}), B = \mathcal{P}(\mathbb{Z})$ (use Problem 57 below)

28. $A = \mathcal{P}\big(\mathcal{P}(\mathbb{N})\big), B = \mathcal{P}(\mathbb{Q})$ (use Problem 57 below)

LEVEL 3

Let $f \colon \mathbb{N} \to \mathbb{Z}$ be defined by $f(n) = \begin{cases} \dfrac{n}{2} & \text{if } n \text{ is even.} \\ -\dfrac{n+1}{2} & \text{if } n \text{ is odd.} \end{cases}$

29. Show that f is injective.

30. Show that f is surjective.

31. Show that f is bijective.

Let $a, b \in \mathbb{R}$ with $a \neq 0$ and $a < b$. Define $f: [0, 1] \to [a, b]$ by $f(x) = (b - a)x + a$.

32. Show that $f(0) = a$.

33. Show that $f(1) = b$.

34. Show that for any x with $0 < x < 1$, we have $a < f(x) < b$.

35. Show that f is injective.

36. Show that f is surjective.

37. Show that f is bijective.

The functions $f: \mathbb{R} \to (0, \infty)$ and $g: (0, \infty) \to (0, 1)$ defined by $f(x) = 2^x$ and $g(x) = \frac{1}{x^2+1}$ are bijections. Use this information to verify each of the following.

38. $\mathbb{R} \sim (0, \infty)$.

39. $(0, \infty) \sim (0, 1)$.

40. $(0, 1) \sim \mathbb{R}$.

41. $[0, 1] \sim \mathbb{R}$.

42. $(0, 1] \sim \mathbb{R}$.

43. $[0, 1) \sim \mathbb{R}$.

44. $(-\infty, 1) \sim (0, \infty)$.

LEVEL 4

Show that each of the following is true.

45. If $b > 0$, then $[0, 1] \sim [0, b]$.

46. If $a \neq 0$ and $a < b$, then $[0, 1] \sim [a, b]$.

47. Any two bounded closed intervals are equinumerous.

48. Any two bounded intervals are equinumerous.

49. Any two intervals are equinumerous (including $\mathbb{R}$ itself).

Show that each of the following is true.

50. $^{\mathbb{N}}\mathbb{N} \preccurlyeq \mathcal{P}(\mathbb{N})$. (you may use Problems 53 and 57 below)

51. $\mathcal{P}(\mathbb{N}) \preccurlyeq {}^{\mathbb{N}}\mathbb{N}$. (you may use Problem 53 below)

52. $^{\mathbb{N}}\mathbb{N} \sim \mathcal{P}(\mathbb{N})$.

Let A, B, C, and D be sets. Show that each of the following is true.

53. $\preccurlyeq$ is transitive.

54. $\prec$ is transitive.

55. If $A \preccurlyeq B$ and $B \prec C$, then $A \prec C$.

56. If $A \prec B$ and $B \preccurlyeq C$, then $A \prec C$.

57. If $A \sim B$, then $\mathcal{P}(A) \sim \mathcal{P}(B)$.

58. If $A \sim B$ and $C \sim D$, then $A \times C \sim B \times D$.

LEVEL 5

Define $\mathcal{P}_k(\mathbb{N})$ for each $k \in \mathbb{N}$ by $\mathcal{P}_0(\mathbb{N}) = \mathbb{N}$ and $\mathcal{P}_{k+1}(\mathbb{N}) = \mathcal{P}\big(\mathcal{P}_k(\mathbb{N})\big)$ for $k > 0$.

59. Explain why $\mathcal{P}_k(\mathbb{N}) \prec \mathcal{P}_{k+1}(\mathbb{N})$ for each $k \in \mathbb{N}$.

60. Find a set B such that for all $k \in \mathbb{N}$, $\mathcal{P}_k(\mathbb{N}) \prec B$.

61. Find a set C such that $B \prec C$.

62. Is there a set X such that $A \preccurlyeq X$ for all sets A?

Let A, B, C, and D be sets. Show that each of the following is true.

63. If $A \sim B$ and $C \sim D$, then $^A C \sim {}^B D$.

64. $^{B \times C}A \sim {}^C(^B A)$.

Let $\{A_n \mid n \in \mathbb{N}\}$ be a pairwise disjoint collection of countable sets and let $\boldsymbol{P} = \{P_n \mid n \in \mathbb{N}\}$ be a partition of $\mathbb{N}$ such that each P_n is infinite (this exists by Problem 80 in Problem Set 4).

65. Explain why there is an injection $f_n : A_n \to P_n$ for each $n \in \mathbb{N}$.

66. If $f : \bigcup\{A_n \mid n \in \mathbb{N}\} \to \mathbb{N}$ is defined by $f(x) = f_n(x)$ if $x \in A_n$, explain why f is well-defined (where f_n was described in Problem 65).

67. Show that f (as defined in Problem 66) is injective.

68. Show that a countable union of countable sets is countable.

CHALLENGE PROBLEMS

69. Show that $\mathbb{R}$ is uncountable.

70. Show that $\mathcal{P}(\mathbb{N}) \sim \{f \in {}^{\mathbb{N}}\mathbb{N} \mid f \text{ is a bijection}\}$.

71. Show that ${}^{\mathbb{N}}\mathbb{R}$ and ${}^{\mathbb{R}}\mathbb{N}$ are **not** equinumerous.

72. Explain why the Cantor-Schroeder-Bernstein Theorem (Equinumerosity Fact 3) is true.

LESSON 7
LOGIC AND AXIOMS

Statements

In mathematics, a **statement** (or **proposition**) is a sentence that can be true or false, but not both simultaneously.

Example 7.1: "Dennis is hungry" is a statement because at any given time either Dennis is hungry or Dennis is not hungry.

Example 7.2: The sentence "Stop doing that!" is *not* a statement because it cannot be true or false. This sentence is a **command**.

Exercise 7.3: Determine if each of the following sentences are statements:

1. Pigs have wings. _____
2. What is wrong with you? _____
3. Dinosaurs are extinct. _____
4. Please read the fine print. _____
5. Andrea is not feeling well today. _____

An **atomic statement** expresses a single idea. The statement "Dennis is hungry" that we discussed above is an example of an atomic statement. Let's look at a few more examples.

Example 7.4: The following sentences are atomic statements:

1. 5 is a prime number.
2. A giraffe is a mammal.
3. $7 > 8$.
4. Jacob weighs 165 pounds.
5. There is more than one universe.

Notes: Sentences 1 and 2 above are true atomic statements and sentence 3 is a false atomic statement.

We can't say for certain whether sentence 4 is true or false without knowing who Jacob is. However, it is either true or false. Therefore, it is a statement. Since it expresses a single idea, it is an atomic statement.

It is also unknown whether sentence 5 is true or false, but this does not change the fact that it must be either true or false. Furthermore, it expresses a single idea. Therefore, it is an atomic statement.

We use **logical connectives** to form **compound statements**. The most commonly used logical connectives are "and," "or," "if...then," "if and only if," and "not."

Example 7.5: The following sentences are compound statements:

1. 5 is a prime number and $0 = 1$.

2. Daniella is dancing or sharks have teeth.

3. If Jimmy is a green parrot, then pigs can fly.

4. Jupiter is smaller than Earth if and only if $1 + 2 = 3$.

5. 9 is not an odd number.

Sentence 1 above uses the logical connective "and." Since the statement "$0 = 1$" is false, it follows that sentence 1 is false. It does not matter that the statement "5 is a prime number" is true. In fact, "T and F" is always F.

Sentence 2 uses the logical connective "or." Since the statement "sharks have teeth" is true, it follows that sentence 2 is true. It does not matter whether Daniella is dancing. In fact, "T or T" is always true and "F or T" is always T.

It's worth pausing for a moment to note that in the English language the word "or" has two possible meanings. There is an "inclusive or" and an "exclusive or." The "inclusive or" is true when both statements are true, whereas the "exclusive or" is false when both statements are true. In mathematics, by default, we always use the "inclusive or" unless we are told to do otherwise. To some extent, this is an arbitrary choice that mathematicians have agreed upon. However, it can be argued that it is the better choice since it is used more often and it is easier to work with. Note that we were assuming use of the "inclusive or" in the last paragraph when we said, "In fact, "T or T" is always true." See Problems 27 and 28 below for more on the "exclusive or."

Sentence 3 uses the logical connective "if...then." The statement "pigs can fly" is false. We need to know whether Jimmy is a green parrot in order to figure out the truth value of sentence 3. If Jimmy is a green parrot, then sentence 3 is false ("if T, then F" is always F). If Jimmy is not a green parrot, then sentence 3 is true ("if F, then F" is always T). Do not worry if you are confused about where the truth values just mentioned come from. We will discuss the logical connective "if...then" (as well as all the other connectives) in much more detail in the section on logical connectives below.

Sentence 4 uses the logical connective "if and only if." Since the two atomic statements have different truth values, it follows that sentence 4 is false. In fact, "F if and only if T" is always F.

Sentence 5 uses the logical connective "not." Since the statement "9 is an odd number" is true, it follows that sentence 5 is false. In fact, "not T" is always F.

Notes: (1) The logical connectives "and," "or," "if...then," and "if and only if," are called **binary connectives** because they join two statements (the prefix "bi" means "two").

(2) The logical connective "not" is called a **unary connective** because it is applied to just a single statement ("unary" means "acting on a single element").

106

(3) Don't worry if the meaning of any of these logical connectives confuses you. We will learn more about them in the section on logical connectives below.

Exercise 7.6: Determine if each of the following statements is an atomic statement or a compound statement.

1. Grace is tired. _____

2. They are flying or they are swimming. _____

3. Dr. Chang did not go to work today. _____

4. I read the book *Crime and Punishment*. _____

5. If pigs can fly, then I will try chicken liver. _____

6. In the beginning of everything, there was only darkness. _____

7. You must set your alarm or you will not wake up. _____

8. I will agree to these terms if and only if my lawyer allows me to agree to them. _____

9. The word "not" has a negative connotation. _____

10. Sarah can play the piano, but she cannot play the violin. _____

Example 7.7: The following sentences are **not** statements:

1. Why do you do this to yourself?

2. Shut the door.

3. $x - 7 = 12$

4. This sentence is false.

5. This sentence is true.

Sentence 1 above is a question and sentence 2 is a command.

Sentence 3 has an unknown variable – it can be turned into a statement by assigning a value to the variable.

Sentences 4 and 5 are self-referential (they refer to themselves). They can be neither true nor false. Sentence 4 is called the Liar's paradox and sentence 5 is called a vacuous affirmation.

Truth Assignments

We will use letters such as p, q, r, and s to denote atomic statements. We will refer to these letters as **propositional variables**, and we will generally assign a truth value of T (for true) or F (for false) to each propositional variable. Formally, we define a **truth assignment** of a list of propositional variables to be a choice of T or F for each propositional variable in the list.

Example 7.8: Consider the propositional variable p. There are **two** possible truth assignments for this propositional variable as follows:

1. We can assign p to be true.
2. We can assign p to be false.

We can visualize this list of truth assignments with the following table:

p
T
F

Observe how the table has just one column because there is only one propositional variable. We label the column with the propositional variable p. Underneath the propositional variable, we have two rows—one for each of the two possible truth assignments.

Example 7.9: Consider the propositional variables p and q (where p and q are **distinct**). There are **four** possible truth assignments for this list of propositional variables as follows:

1. We can assign both p and q to be true.
2. We can assign p to be true and q to be false.
3. We can assign p to be false and q to be true.
4. We can assign both p and q to be false.

Note: When we say that p and q are **distinct**, we mean that $p \neq q$. In other words, we use the word distinct when we want to make sure that it is understood that the objects under consideration are different from each other.

We can visualize this list of truth assignments with the following table:

p	q
T	T
T	F
F	T
F	F

Observe how the table has two columns because there are two propositional variables. We label the columns with the propositional variables p and q. Underneath the propositional variables, we have four rows—one for each of the four possible truth assignments.

Exercise 7.10: Consider the three distinct propositional variables p, q, and r. How many different truth assignments are there for this list of propositional variables? _____

Draw a table that will allow us to visualize this list of truth assignments.

Logical Connectives

We use the symbols ∧, ∨, →, ↔, and ¬ for the most common logical connectives. The truth value of a compound statement is determined by the truth values of its atomic parts together with applying various rules for the connectives. Let's look at each connective in detail.

We will use the "wedge" symbol ∧ to represent the logical connective "and." The compound statement $p \wedge q$ is called the **conjunction** of p and q. It is pronounced "**p and q**." $p \wedge q$ is true when both p and q are true, and it is false otherwise.

The following table summarizes the truth values of $p \wedge q$ for each possible truth assignment of the propositional variables p and q.

p	q	$p \wedge q$
T	T	T
T	F	F
F	T	F
F	F	F

Notes: (1) The table displayed above is called a **truth table**. This type of table is used to display the possible truth values of a compound statement. We start by labelling the columns of the table with the propositional variables that appear in the statement, followed by the statement itself. We then use the rows to run through every possible combination of truth values for the propositional variables (all possible truth assignments) followed by the resulting truth values for the compound statement.

(2) The first two columns of the truth table above (labeled p and q) give the four possible truth assignments for the propositional variables p and q (see Example 7.9).

(3) The truth table for the conjunction is based upon the way we use the word "and" in everyday English. For example, suppose that Jamie is a girl with black hair. Then the statement "Jamie is a girl and Jamie has black hair" is true because each of the statements "Jamie is a girl" and "Jamie has black hair" is true. Similarly, the statement "Jamie is a girl and Jamie has red hair" is false because the statement "Jamie has red hair" is false. Based upon your own experience of the English language, you may be able to compute the truth value of a conjunction of two statements without needing to look back at the truth table.

Example 7.11: If p is true and q is false, then we can compute the truth value of $p \wedge q$ by looking at the second row of the truth table for the conjunction.

p	q	$p \wedge q$
T	T	T
T	F	F
F	T	F
F	F	F

We see from the highlighted row in the truth table above that $p \wedge q \equiv T \wedge F \equiv \textbf{F}$.

Note: Here the symbol $\equiv$ can be read "is logically equivalent to." So, we see that if p is true and q is false, then $p \wedge q$ is logically equivalent to F, or more simply, $p \wedge q$ is false.

Exercise 7.12: Determine the truth value of $p \wedge q$ given that

1. p and q are both true. ____

2. p and q are both false. ____

3. p is false and q is true. ____

We will use the "vee" symbol $\vee$ to represent the logical connective "or." The compound statement $p \vee q$ is called the **disjunction** of p and q. It is pronounced "**p or q**." $p \vee q$ is true when p or q (or both) are true, and it is false when p and q are both false.

The following table summarizes the truth values of $p \vee q$ for each possible truth assignment of the propositional variables p and q.

p	q	$p \vee q$
T	T	T
T	F	T
F	T	T
F	F	F

Notes: (1) The truth table for the disjunction is based upon the way we use the "inclusive or" in everyday English. For example, suppose that in order to be able to watch television, Kelly's parents tell her that she must first either do the dishes or clean her room. In this context, the statement "Kelly does the dishes or Kelly cleans her room" is understood to be true if Kelly does the dishes or Kelly cleans her room **or both**. Certainly if she completes both of these tasks, it would be unnatural to penalize her.

(2) In mathematics, when we use the word "or" we always assume that we mean the "inclusive or" unless we are told otherwise. In English when we use the word "or," we are more likely to be using the "exclusive or." For example, if a waiter says "you can have fries or a salad with your order," it is unlikely that he means you can have both. Indeed, this is an example of the "exclusive or." See Problems 27 and 28 below for more on the exclusive or.

Exercise 7.13: Determine the truth value of $p \vee q$ given that

1. p and q are both true. ____

2. p and q are both false. ____

3. p is true and q is false. ____

4. p is false and q is true. ____

We will use the "rightarrow" symbol $\rightarrow$ to represent the logical connective "if...then." The compound statement $p \rightarrow q$ is called a **conditional** or **implication**. It is pronounced "**if p, then q**" or **p implies q**. $p \rightarrow q$ is true when p is false or q is true (or both), and it is false when p is true and q is false.

The following table summarizes the truth values of $p \rightarrow q$ for each possible truth assignment of the propositional variables p and q.

p	q	$p \rightarrow q$
T	T	T
T	F	F
F	T	T
F	F	T

Notes: (1) In the conditional $p \rightarrow q$, p is called the **hypothesis** (or **assumption** or **premise**) and q is called the **conclusion**.

(2) The truth table for the conditional is loosely based upon the English meaning of "if...then" or "implies." However, trying to understand this truth table by analyzing English sentences often leads to confusion. Therefore, in this case, it may be more instructive to understand what we wish to accomplish with this connective mathematically. When should $p \rightarrow q$ be true. Well, we would like $p \rightarrow q$ to be true if the assumption that p is true forces q to be true as well (equivalently, if the hypothesis is true, then the conclusion must be true). For example, let's take the statement "If Odin is a cat, then Odin can bark." Now, if Odin happens to be a cat, then the statement just given in quotes is false. Do you see why? The hypothesis "Odin is a cat" is true. If the conditional were true, then the conclusion "Odin can bark" would be forced to be true. However, cats can't bark. So, the conditional is false. This situation corresponds to the second row in the truth table for the conditional above.

On the other hand, the statement "If Odin is a cat, then Odin can meow" is most likely true. This time the hypothesis and conclusion are both true. This situation corresponds to the first row in the truth table for the conditional above.

(3) What about the situation in which the hypothesis is false. In this case, we don't really care what the conclusion is. The conditional is true either way. When the hypothesis is false, we will say that the conditional statement is **vacuously true**. The word "vacuous" means "empty." The idea is that something that is vacuously true is true for a very silly reason. For example, suppose that we are looking at an empty room and someone says, "If there is a pig in that room, then it can fly." This is equivalent to saying, "Every pig in that room can fly." This statement is true, but for a very dumb reason. Yes, every pig in that room can fly simply because there are no pigs in the room. If there were even a single pig in the room, then the statement would be false. In order for someone to dispute our claim that every pig in the room can fly, they would need to show us a pig in the room that cannot fly. Of course, they cannot do this. After all, there are no pigs in the room. This notion of "vacuous truth" corresponds to the third and fourth rows in the truth table for the conditional above.

Exercise 7.14: Determine the truth value of $p \rightarrow q$ given that

1. p and q are both true. ____

2. p and q are both false. ____

3. p is true and q is false. ____

4. p is false and q is true. ____

We will use the "doublearrow" symbol $\leftrightarrow$ to represent the logical connective "if and only if." The compound statement $p \leftrightarrow q$ is called a **biconditional**. It is pronounced "**p if and only if q**." $p \leftrightarrow q$ is true when p and q have the same truth value (both true or both false), and it is false when p and q have opposite truth values (one true and the other false).

The following table summarizes the truth values of $p \leftrightarrow q$ for each possible truth assignment of the propositional variables p and q.

p	q	$p \leftrightarrow q$
T	T	T
T	F	F
F	T	F
F	F	T

Exercise 7.15: Determine the truth value of $p \leftrightarrow q$ given that

1. p and q are both true. ____

2. p and q are both false. ____

3. p is true and q is false. ____

4. p is false and q is true. ____

We will use the "taildash" symbol ¬ to represent the logical connective "not." The compound statement ¬p is called the **negation** of p. It is pronounced "**not** p." ¬p is true when p is false, and it is false when p is true (p and ¬p have opposite truth values).

The following table summarizes the truth values of ¬p for each possible truth assignment of the propositional variable p.

p	¬p
T	F
F	T

Notes: (1) Since negation requires only a single propositional variable, there are just two possible truth assignments to worry about. The first column of the truth table above (labeled p) gives the two possible truth assignments (T and F).

(2) The truth table for the negation is based upon the way we use the word "not" in everyday English. For example, since the statement "Fish swim" is true, it follows that the statement "Fish do not swim" is false. Similarly, since the statement "Elephants fly" is false, it follows that the statement "Elephants do not fly" is true.

Example 7.16: If p is true, then we can compute the truth value of ¬p by looking at the first row of the truth table for the negation.

p	¬p
T	F
F	T

We see from the highlighted row in the truth table above that ¬p ≡ ¬T ≡ **F**.

Exercise 7.17: Determine the truth value of ¬p given that p is false. ___

Example 7.18: Let p represent the statement "Ducks quack" and let q represent the statement "$0 = 1$." Note that p is true and q is false.

1. $p \wedge q$ represents "Ducks quack and $0 = 1$." Since q is false, it follows that $p \wedge q$ is false.

2. $p \vee q$ represents "Ducks quack or $0 = 1$." Since p is true, it follows that $p \vee q$ is true.

3. $p \rightarrow q$ represents "If ducks quack, then $0 = 1$." Since p is true and q is false, $p \rightarrow q$ is false.

4. $p \leftrightarrow q$ represents "Ducks quack if and only if $0 = 1$." Since p is true and q is false, $p \leftrightarrow q$ is false.

5. ¬q represents the statement "$0 \neq 1$." Since q is false, ¬q is true.

6. ¬$p \vee q$ represents the statement "Ducks don't quack or $0 = 1$." Since ¬p and q are both false, ¬$p \vee q$ is false. Note that ¬$p \vee q$ always means $(¬p) \vee q$. In general, without parentheses present, we always apply negation before any of the other connectives.

113

7. $\neg(p \lor q)$ represents the statement "It is not the case that either ducks quack or $0 = 1$." This can also be stated as "Neither do ducks quack nor is 0 equal to 1." Since $p \lor q$ is true (see 2 above), $\neg(p \lor q)$ is false.

8. $\neg p \land \neg q$ represents the statement "Ducks don't quack and $0 \neq 1$." This statement can also be stated as "Neither do ducks quack nor is 0 equal to 1." Since this is the same statement as in 7 above, it should follow that $\neg p \land \neg q$ is equivalent to $\neg(p \lor q)$. You will be asked to verify this later (see Exercise 7.27 below). For now, let's observe that since $\neg p$ is false, it follows that $\neg p \land \neg q$ is false. This agrees with the truth value we got in 7.

Note: The equivalence of $\neg(p \lor q)$ with $\neg p \land \neg q$ (see parts 7 and 8 of Example 7.18 above) is one of **De Morgan's laws**. These laws will be explored further below (see Example 7.26 and Exercise 7.27).

Exercise 7.19: Let p represent the statement "Frogs are birds," and let q represent the statement "$2 < 1$." Translate each of the following compound statements into English and compute the truth value of each statement.

1. $p \rightarrow q$ _____

2. $\neg p \lor q$ _____

3. $p \leftrightarrow q$ _____

4. $(p \rightarrow q) \land (q \rightarrow p)$ _____

5. $\neg(p \land q)$ _____

6. $\neg p \lor \neg q$ _____

Evaluating Truth

Example 7.20: Let p, q, and r be propositional variables with p and q true, and r false. Let's compute the truth value of $\neg p \lor (\neg q \rightarrow r)$.

Truth table solution: One foolproof way to compute the desired truth value is to build the whole truth table of $\neg p \lor (\neg q \rightarrow r)$ one column at a time. Since there are 3 propositional variables (p, q, and r), we will need 8 rows to get all the possible truth values (see Exercise 7.10 and its solution). We then create a column for each compound statement that appears within the given statement starting with the statements of smallest length and working our way up to the given statement. We will need columns for p, q, r (the atomic statements), $\neg p$, $\neg q$, $\neg q \rightarrow r$, and finally, the statement itself, $\neg p \lor (\neg q \rightarrow r)$. Below is the final truth table with the relevant row highlighted and the final answer circled.

p	q	r	$\neg p$	$\neg q$	$\neg q \to r$	$\neg p \vee (\neg q \to r)$
T	T	T	F	F	T	T
T	T	F	F	F	T	Ⓣ
T	F	T	F	T	T	T
T	F	F	F	T	F	F
F	T	T	T	F	T	T
F	T	F	T	F	T	T
F	F	T	T	T	T	T
F	F	F	T	T	F	T

Notes: (1) We fill out the first three columns of the truth table by listing all possible combinations of truth assignments for the propositional variables p, q, and r. Notice how down the first column we have 4 T's followed by 4 F's, down the second column we alternate sequences of 2 T's with 2 F's, and down the third column we alternate T's with F's one at a time. This is a nice systematic way to make sure we get all possible combinations of truth assignments.

If you're having trouble seeing the pattern of T's and F's, here is another way to think about it: In the first column, the first half of the rows have a T and the remainder have an F. This gives 4 T's followed by 4 F's.

For the second column, we take half the number of consecutive T's in the first column (half of 4 is 2) and then we alternate between 2 T's and 2 F's until we fill out the column.

For the third column, we take half the number of consecutive T's in the second column (half of 2 is 1) and then we alternate between 1 T and 1 F until we fill out the column.

(2) Since the connective $\neg$ has the effect of taking the opposite truth value, we generate the entries in the fourth column by taking the opposite of each truth value in the first column. Similarly, we generate the entries in the fifth column by taking the opposite of each truth value in the second column.

(3) For the sixth column, we apply the connective $\to$ to the fifth and third columns, respectively, and finally, for the last column, we apply the connective $\vee$ to the fourth and sixth columns, respectively.

(4) The original question is asking us to compute the truth value of $\neg p \vee (\neg q \to r)$ when p and q are true, and r is false. In terms of the truth table, we are being asked for the entry in the second row and last (seventh) column. Therefore, the answer is **T**.

(5) This is certainly not the most efficient way to answer the given question. However, building truth tables is not too difficult, and it's a foolproof way to determine truth values of compound statements.

Alternate Solution: We have $\neg p \vee (\neg q \rightarrow r) \equiv \neg T \vee (\neg T \rightarrow F) \equiv F \vee (F \rightarrow F) \equiv F \vee T \equiv \mathbf{T}.$

Notes: (1) For the first equivalence, we simply replaced the propositional variables by their given truth values. We replaced p and q by T, and we replaced r by F.

(2) For the second equivalence, we used the first row of the truth table for the negation (drawn to the right for your convenience).

p	$\neg p$
T	F
F	T

We see from the highlighted row that $\neg T \equiv F$. We applied this result twice.

(3) For the third equivalence, we used the fourth row of the truth table for the conditional.

p	q	$p \rightarrow q$
T	T	T
T	F	F
F	T	T
F	F	T

We see from the highlighted row that $F \rightarrow F \equiv T$.

(4) For the last equivalence, we used the third row of the truth table for the disjunction.

p	q	$p \vee q$
T	T	T
T	F	T
F	T	T
F	F	F

We see from the highlighted row that $F \vee T \equiv T$.

(5) We can save a little time by immediately replacing the negation of a propositional variable by its truth value (which will be the opposite truth value of the propositional variable). For example, since p has truth value T, we can replace $\neg p$ by F. Similarly, since q has truth value T, we can replace $\neg q$ by F. The faster solution would look like this:

$$\neg p \vee (\neg q \rightarrow r) \equiv F \vee (F \rightarrow F) \equiv F \vee T \equiv \mathbf{T}.$$

Quicker solution: Since q has truth value T, it follows that $\neg q$ has truth value F. So, $\neg q \rightarrow r$ has truth value T. Finally, $\neg p \vee (\neg q \rightarrow r)$ must then have truth value T.

Notes: (1) Symbolically, we can write the following:

$$\neg p \vee (\neg q \rightarrow r) \equiv \neg p \vee (\neg T \rightarrow r) \equiv \neg p \vee (F \rightarrow r) \equiv \neg p \vee T \equiv \mathbf{T}$$

(2) We can display this reasoning visually as follows:

$$\neg p \vee (\neg q \rightarrow r)$$

$$
\begin{array}{ccc}
 & T & \\
 & F & \\
 & & T \\
T & &
\end{array}
$$

The vertical lines have just been included to make sure you see which connective each truth value is written below.

We began by placing a T under the propositional variable q to indicate that q is true. Since $\neg T \equiv F$, we then place an F under the negation symbol. Next, since $F \rightarrow r \equiv T$ regardless of the truth value of r, we place a T under the conditional symbol. Finally, since $\neg p \vee T \equiv T$ regardless of the truth value of p, we place a T under the disjunction symbol. We made this last T bold to indicate that we are finished.

(3) Knowing that q has truth value T is enough to determine the truth value of $\neg p \vee (\neg q \rightarrow r)$, as we saw in Note 1 above. It's okay if you didn't notice that right away. This kind of reasoning takes a bit of practice and experience.

Exercise 7.21: Let p, q, and r be propositional variables.

1. Draw the truth table for $p \leftrightarrow (q \wedge \neg r)$.

2. Use the truth table from part 1 to compute the truth value of $p \leftrightarrow (q \wedge \neg r)$ when p is true, and q and r are false. _____

3. Suppose that p and r are both true. Is this enough information to compute the truth value of $p \leftrightarrow (q \wedge \neg r)$? _____ If so, what is that truth value? _____

Logical Equivalence

We say that two statements are **logically equivalent** if they have the same truth table. We use the symbol "$\equiv$" to indicate logical equivalence.

Example 7.22: Let p be a propositional variable. Let's show that p and $\neg(\neg p)$ are logically equivalent (symbolically, we write $p \equiv \neg(\neg p)$). We will show that p and $\neg(\neg p)$ have the same truth table. We can put all the information into a single table.

p	$\neg p$	$\neg(\neg p)$
T	F	T
F	T	F

Observe that the first column gives the truth values for p, the third column gives the truth values for $\neg(\neg p)$, and both these columns are identical. It follows that $p \equiv \neg(\neg p)$.

Notes: (1) The logical equivalence $p \equiv \neg(\neg p)$ is called the **law of double negation**. In words, this law says that if you negate a propositional variable twice, the resulting statement is logically equivalent to the original propositional variable.

(2) As an example in English, let p be the statement "John is hungry." Then the statement $\neg(\neg p)$ can be expressed in English as "It is not the case that John is not hungry." By the law of double negation, these two statements are logically equivalent.

Exercise 7.23: The **law of the conditional** is the logical equivalence $p \to q \equiv \neg p \vee q$. Use a truth table to verify this logical equivalence.

Note: The law of the conditional allows us to replace the conditional statement $p \to q$ by the more intuitive statement $\neg p \vee q$. Remember that we can think of the conditional statement $p \to q$ as having the hypothesis p and the conclusion q. The disjunctive form $\neg p \vee q$ tells us quite explicitly that a conditional statement is true precisely if the hypothesis p is false or the conclusion q is true (or both).

Consider the conditional statement $p \to q$. There are three other statements associated with this statement.

1. The **converse** is the statement $q \to p$.

2. The **inverse** is the statement $\neg p \to \neg q$.

3. The **contrapositive** is the statement $\neg q \to \neg p$.

Example 7.24: Consider the conditional statement "If you are a cat, then you are a mammal."

1. The converse is the statement "If you are a mammal, then you are a cat."

2. The inverse is the statement "If you are not a cat, then you are not a mammal."

3. The contrapositive is the statement "If you are not a mammal, then you are not a cat."

Notes: (1) If we let p be the statement "You are a cat" and we let q be the statement "You are a mammal," then the given conditional statement can be represented by $p \to q$. Similarly, the converse can be represented by $q \to p$, the inverse can be represented by $\neg p \to \neg q$, and the contrapositive can be represented by $\neg q \to \neg p$.

(2) Notice that in this example, the given conditional statement is true. Indeed, every cat is a mammal. On the other hand, the converse is false. After all, there are certainly mammals that are not cats. For example, a dog is a mammal that is not a cat. This example shows that a conditional statement is **not** logically equivalent to its converse.

(3) In this example, the inverse is also false. For example, a dog is not a cat, and yet, a dog is a mammal. This example shows that a conditional statement is **not** logically equivalent to its inverse.

(4) In this example, the contrapositive is true. Anything that is not a mammal cannot possibly be a cat. In fact, it turns out that a conditional statement is **always** logically equivalent to its contrapositive. We will see this in Example 7.25 below. A word of caution is in order here. The one example just given does **not** prove that the given conditional statement is logically equivalent to its contrapositive. To verify logical equivalence, we need to check the whole truth table.

Exercise 7.25: The **law of the contrapositive** is the logical equivalence $p \to q \equiv \neg q \to \neg p$. Use a truth table to verify this logical equivalence.

The **De Morgan's laws** provide formulas for negating a conjunction and for negating a disjunction.

$$\neg(p \land q) \equiv \neg p \lor \neg q \qquad \qquad \neg(p \lor q) \equiv \neg p \land \neg q$$

Example 7.26: Let's verify the first De Morgan's law. In other words, we will show that $\neg(p \land q)$ and $\neg p \lor \neg q$ are logically equivalent. We will provide two different methods.

Truth table method:

p	q	$\neg p$	$\neg q$	$p \land q$	$\neg(p \land q)$	$\neg p \lor \neg q$
T	T	F	F	T	F	F
T	F	F	T	F	T	T
F	T	T	F	F	T	T
F	F	T	T	F	T	T

Observe that the sixth column gives the truth values for $\neg(p \wedge q)$, the seventh column gives the truth values for $\neg p \vee \neg q$, and both these columns are identical. It follows that $\neg(p \wedge q) \equiv \neg p \vee \neg q$.

Direct method: If $p \equiv F$ or $q \equiv F$, then $\neg(p \wedge q) \equiv \neg F \equiv T$ and $\neg p \vee \neg q \equiv T$ (because $\neg p \equiv T$ or $\neg q \equiv T$). If $p \equiv T$ and $q \equiv T$, then $\neg(p \wedge q) \equiv \neg T \equiv F$ and $\neg p \vee \neg q \equiv F \vee F \equiv F$. So, all four possible truth assignments of p and q lead to the same truth value for $\neg(p \wedge q)$ and $\neg p \vee \neg q$. It follows that $\neg(p \wedge q) \equiv \neg p \vee \neg q$.

Exercise 7.27: Use a truth table to verify the second De Morgan's law $\neg(p \vee q) \equiv \neg p \wedge \neg q$.

List 7.28: Here is a list of some useful logical equivalences. The dedicated reader may want to verify each of these by drawing a truth table or by using direct arguments similar to that used in Example 7.26 (some of these are asked for in Problems 63 through 68 below).

1. **Law of double negation:** $p \equiv \neg(\neg p)$

2. **De Morgan's laws:** $\quad \neg(p \wedge q) \equiv \neg p \vee \neg q \qquad\qquad \neg(p \vee q) \equiv \neg p \wedge \neg q$

3. **Commutative laws:** $\quad p \wedge q \equiv q \wedge p \qquad\qquad\quad p \vee q \equiv q \vee p$

4. **Associative laws:** $\quad (p \wedge q) \wedge r \equiv p \wedge (q \wedge r) \qquad (p \vee q) \vee r \equiv p \vee (q \vee r)$

5. **Distributive laws:** $\quad p \wedge (q \vee r) \equiv (p \wedge q) \vee (p \wedge r) \qquad p \vee (q \wedge r) \equiv (p \vee q) \wedge (p \vee r)$

6. **Identity laws:** $\quad\quad p \wedge T \equiv p \qquad p \wedge F \equiv F \qquad p \vee T \equiv T \qquad p \vee F \equiv p$

7. **Negation laws:** $\quad\quad p \wedge \neg p \equiv F \qquad\qquad\qquad p \vee \neg p \equiv T$

8. **Redundancy laws:** $\quad p \wedge p \equiv p \qquad\qquad\qquad p \vee p \equiv p$

9. **Absorption laws:** $\quad (p \vee q) \wedge p \equiv p \qquad\qquad (p \wedge q) \vee p \equiv p$

10. **Law of the conditional:** $\quad p \rightarrow q \equiv \neg p \vee q$

11. **Law of the contrapositive:** $p \rightarrow q \equiv \neg q \rightarrow \neg p$

12. **Law of the biconditional:** $\quad p \leftrightarrow q \equiv (p \rightarrow q) \wedge (q \rightarrow p)$

Note: Although this is a fairly long list of laws, a lot of it is quite intuitive. For example, in English the word "and" is commutative. The statements "I have a cat and I have a dog" and "I have a dog and I have a cat" have the same meaning. So, it's easy to see that $p \wedge q \equiv q \wedge p$ (the first law in 3 above). As another example, the statement "I have a cat and I do not have a cat" could never be true. So, it's easy to see that $p \wedge \neg p \equiv F$ (the first law in 7 above).

Example 7.29: Let's show that the statement $p \wedge [(p \wedge \neg q) \vee q]$ is logically equivalent to the atomic statement p.

Solution:
$$p \wedge [(p \wedge \neg q) \vee q] \equiv p \wedge [q \vee (p \wedge \neg q)] \equiv p \wedge [(q \vee p) \wedge (q \vee \neg q)] \equiv p \wedge [(q \vee p) \wedge T]$$
$$\equiv p \wedge (q \vee p) \equiv (q \vee p) \wedge p \equiv (p \vee q) \wedge p \equiv p$$

So, we see that $p \wedge [(p \wedge \neg q) \vee q]$ is logically equivalent to the atomic statement p.

Notes: (1) For the first equivalence, we used the second commutative law.

(2) For the second equivalence, we used the second distributive law.

(3) For the third equivalence, we used the second negation law.

(4) For the fourth equivalence, we used the first identity law.

(5) For the fifth equivalence, we used the first commutative law.

(6) For the sixth equivalence, we used the second commutative law.

(7) For the last equivalence, we used the first absorption law.

Exercise 7.30: Show that the statement $[(\neg p \vee q) \wedge p] \vee q$ is logically equivalent to q.

_____

_____

_____

Tautologies and Contradictions

A statement that has truth value T for all truth assignments of the propositional variables is called a **tautology**. Similarly, a statement that has truth value F for all truth assignments of the propositional variables is called a **contradiction**.

Example 7.31: Let's show that the statement $p \rightarrow p$ is a tautology.

Direct method: If $p \equiv T$, then $p \rightarrow p \equiv T \rightarrow T \equiv T$. If $p \equiv F$, then $p \rightarrow p \equiv F \rightarrow F \equiv T$. Since both possible truth assignments of the propositional variable p lead to the statement $p \rightarrow p$ having truth value T, it follows that $p \rightarrow p$ is a tautology.

Truth table method:

p	$p \rightarrow p$
T	T
F	T

Since the last column of the truth table consists of only the truth value T, the statement $p \rightarrow p$ is a tautology.

Exercise 7.32: Use a truth table to show that $(p \to q) \leftrightarrow (\neg q \to \neg p)$ is a tautology.

Note: Observe the similarity between the tautology $(p \to q) \leftrightarrow (\neg q \to \neg p)$ and the law of the contrapositive $p \to q \equiv \neg q \to \neg p$ (this is logical equivalence 11 from List 7.28). Given any logical equivalence $A \equiv B$ (where A and B are sentences), we always have a corresponding tautology $A \leftrightarrow B$.

Example 7.33: From the first De Morgan's Law $\neg(p \wedge q) \equiv \neg p \vee \neg q$, it follows that the statement $\neg(p \wedge q) \leftrightarrow \neg p \vee \neg q$ is a tautology.

Exercise 7.34: Show directly that the statement $p \wedge \neg p$ is a contradiction (by "directly," we mean that you should **not** use a truth table).

Quantifiers

There are two "quantifiers" that appear repeatedly in both informal and formal mathematics. The **universal quantifier**, written $\forall$, is pronounced "For all," "For every," or "For each." The **existential quantifier**, written $\exists$, is pronounced "For some," "For at least one," or "There exists."

Without adding any restrictions, the statements "For all" and "There exists" may seem way too general. When we say "For all," what exactly do we mean? Do we mean "For all people?" Do we mean "For all complex numbers?" To make this clear, we should always start off with a specific **universe** (or **domain of discourse**), U, that we have decided to work within.

Example 7.35:

1. Let $U = \{x \mid x \text{ is an animal}\}$ be the universe consisting of all animals. Let $C(x)$ represent the statement "x is a cat." Let a represent "Achilles" and let o represent "Odin." Then $C(a)$ represents "Achilles is a cat" and $C(o)$ represents "Odin is a cat." Assuming that Achilles and Odin are both cats, the statements $C(a)$ and $C(o)$ are both true.

   The statement $\forall x\big(C(x)\big)$ can be read "For all x, x is a cat." Since our universe consists of all animals, this reduces to "All animals are cats." Therefore, $\forall x\big(C(x)\big)$ is a false statement.

   Similarly, the statement $\exists x\big(C(x)\big)$ can be read "For some x, x is a cat," or "Some animals are cats." Therefore, $\exists x\big(C(x)\big)$ is a true statement.

2. Define U and $C(x)$ as in part 1 above, and also let $B(x)$ represent the statement "x has black fur." The statement $\forall x\big(C(x) \rightarrow B(x)\big)$ can be read "For all x, if x is a cat, then x has black fur." This can be said more simply as "All cats have black fur." Therefore, $\forall x\big(C(x) \rightarrow B(x)\big)$ is a false statement.

   The statement $\exists x\big(C(x) \wedge B(x)\big)$ can be read "There is an x such that x is a cat and x has black fur. This can be said more simply as "Some cats have black fur." Therefore, $\exists x\big(C(x) \wedge B(x)\big)$ is a true statement.

   Notice how our usage of the **unary relation symbol** C in the statements $\forall x\big(C(x) \rightarrow B(x)\big)$ and $\exists x\big(C(x) \wedge B(x)\big)$ allows us to simulate quantifying over "cats" instead of "animals." In other words, these statements would just be written $\forall x\big(B(x)\big)$ and $\exists x\big(B(x)\big)$, respectively, if the universe was "cats" instead of "animals." This is a neat little trick that allows us to "chop up" a universe in a way that we can quantify over pieces of the universe as needed.

3. Let $U = \mathbb{Z}$, the set of integers, and define $+$ and $<$ in the usual way. The statement $x + y < 0$ is neither true nor false. There are two ways we can modify this statement to give it a truth value. The first way is to substitute values from the universe in for x and y. For example, if we let $x = 3$ and $y = 7$, then we get the false statement $3 + 7 < 0$.

   Another way to modify the statement is to turn the "free variables" x and y into "bound variables" by introducing quantifiers. For example, the statement $\forall x \forall y(x + y < 0)$ is false. Indeed, we just provided a counterexample a moment ago by setting $x = 3$ and $y = 7$. As another example, the statement $\forall x \exists y(x + y < 0)$ is true. Indeed, given x, let $y = -x - 1$. Then $x + y = x + (-x - 1) = -1$ and $-1 < 0$. What about the statement $\exists y \forall x(x + y < 0)$? Given y, the value $x = -y$ provides a counterexample, showing that this statement is false.

   The statements $\forall x \exists y(x + y < 0)$ and $\exists y \forall x(x + y < 0)$ prove that universal quantifiers **do not** commute with existential quantifiers. In other words, in general, the statements $\forall x \exists y\big(P(x, y)\big)$ and $\exists y \forall x\big(P(x, y)\big)$ are **not** logically equivalent.

4. Let $U = \{0, 1\}$ and define $\cdot$ (multiplication) in the usual way. The statement $\forall x(x \cdot x = x)$ is true in this universe. After all, $0 \cdot 0 = 0$ and $1 \cdot 1 = 1$. In this case, the statement $\forall x(x \cdot x = x)$ is logically equivalent to $0 \cdot 0 = 0 \wedge 1 \cdot 1 = 1$.

   More generally, if $U = \{a_1, a_2, \ldots, a_n\}$ is a finite set, then the statement $\forall x\big(P(x)\big)$ is logically equivalent to $P(a_1) \wedge P(a_2) \wedge \cdots \wedge P(a_n)$. Similarly, the statement $\exists x\big(P(x)\big)$ is logically equivalent to $P(a_1) \vee P(a_2) \vee \cdots \vee P(a_n)$.

   In the case where U is an infinite set, a universal quantifier can be thought of as an infinitary conjunction and an existential quantifier can be thought of as an infinitary disjunction. These quantifiers give us a neat little way to talk about all elements of a structure simultaneously even if that structure is infinite.

Exercise 7.36: Let $U = \mathbb{Z}$ and define $+, \cdot, =$, and $\leq$ in the usual way. Determine if each of the following statements is true or false.

1. $\forall x(x \leq x \cdot x)$ _____
2. $\exists x \exists y\big((x + x = x \cdot x) \wedge (y + y = y \cdot y) \wedge x \neq y\big)$ _____
3. $\forall x \forall y(x + y = x \cdot y)$ _____
4. $\forall x \, \exists y(x < y)$ _____
5. $\exists y \forall x(x < y)$ _____

$\neg\forall x\big(P(x)\big)$ is logically equivalent to $\exists x\big(\neg P(x)\big)$. In other words, when we pass a negation symbol through a universal quantifier, the quantifier changes to an existential quantifier. Symbolically, we can write $\neg\forall x\big(P(x)\big) \equiv \exists x\big(\neg P(x)\big)$, where as usual, $\equiv$ is pronounced "is logically equivalent to."

Similarly, $\neg\exists x\big(P(x)\big) \equiv \forall x\big(\neg P(x)\big)$.

Example 7.37:

1. As in part 1 of Example 7.35, let $U = \{x \mid x \text{ is an animal}\}$ and let $C(x)$ represent the statement "x is a cat." Recall that the false statement $\forall x\big(C(x)\big)$ can be read as "All animals are cats." The negation of this statement is the true statement $\neg\forall x\big(C(x)\big)$, which can be read "It is not the case that all animals are cats." By the above remarks, if we simultaneously pass the negation symbol through the universal quantifier and change the universal quantifier to an existential quantifier, we get the logically equivalent statement $\exists x\big(\neg C(x)\big)$, which can be read "Some animals are not cats."

   To summarize the above paragraph, $\neg\forall x\big(C(x)\big) \equiv \exists x\big(\neg C(x)\big)$. In words, "It is not the case that all animals are cats" is logically equivalent to "Some animals are not cats."

   Similarly, the negation of the statement $\exists x\big(C(x)\big)$ is $\neg\exists x\big(C(x)\big)$, which is logically equivalent to $\forall x\big(\neg C(x)\big)$. In words, "It is not the case that there is a cat" is logically equivalent to "Every animal is not a cat." This can be stated a little nicer in English as "No animal is a cat."

2. As in part 2 of Example 7.35, let $B(x)$ represent the statement "x has black fur." Recall that the statement $\forall x\big(C(x) \to B(x)\big)$ can be read "All cats have black fur." The negation of this statement is $\neg\forall x\big(C(x) \to B(x)\big)$, which is logically equivalent to $\exists x\left(\neg\big(C(x) \to B(x)\big)\right)$. Now, by Problem 68 below, $\neg\big(C(x) \to B(x)\big) \equiv C(x) \wedge \neg B(x)$. Therefore, it follows that $\neg\forall x\big(C(x) \to B(x)\big) \equiv \exists x\big(C(x) \wedge \neg B(x)\big)$.

   Let's translate these last two statements into English. $\neg\forall x\big(C(x) \to B(x)\big)$ can be read "It is not the case that all cats have black fur." $\exists x\big(C(x) \wedge \neg B(x)\big)$ can be read "There is a cat that does not have black fur." Notice that both these statements are true and in fact, they have the same meaning in English.

Similarly, the negation of the statement $\exists x\big(C(x) \wedge B(x)\big)$ is $\neg\exists x\big(C(x) \wedge B(x)\big)$, which is logically equivalent to $\forall x\left(\neg\big(C(x) \wedge B(x)\big)\right)$. Now, by De Morgan's law from List 7.28, we have $\neg\big(C(x) \wedge B(x)\big) \equiv \neg C(x) \vee \neg B(x)$. By the law of the conditional from the same list (and the law of double negation), we have $\neg C(x) \vee \neg B(x) \equiv C(x) \rightarrow \neg B(x)$. It follows that $\neg\exists x\big(C(x) \wedge B(x)\big) \equiv \forall x\big(C(x) \rightarrow \neg B(x)\big)$.

Let's translate these last two statements into English. $\neg\exists x\big(C(x) \wedge B(x)\big)$ can be read "It is not the case that some cats have black fur." $\forall x\big(C(x) \rightarrow \neg B(x)\big)$ can be read "Every cat does not have black fur." Notice that both these statements are false and in fact, they have the same meaning in English. Also note that "Every cat does not have black fur" can be stated a little nicer in English as "No cat has black fur."

3. As in part 3 of Example 7.35, let $U = \mathbb{Z}$ and define $+$ and $<$ in the usual way. We have the following sequence of logical equivalences:

$$\neg\forall x\forall y(x + y < 0) \equiv \exists x\big(\neg\forall y(x + y < 0)\big) \equiv \exists x\exists y\big(\neg(x + y < 0)\big) \equiv \exists x\exists y(x + y \geq 0).$$

So, the negation of the statement "For all integers x and y, $x + y$ is negative" is logically equivalent to the statement "There exist integers x and y such that $x + y$ is nonnegative." Notice that the first statement is false, while the second statement is true.

Similarly, $\neg\forall x\exists y(x + y < 0) \equiv \exists x\forall y(x + y \geq 0)$. So, the negation of the statement "For every integer x, there is an integer y such that $x + y$ is negative" is logically equivalent to the statement "There is an integer x such that for every integer y, $x + y$ is nonnegative." Notice that the first statement is true, while the second statement is false.

Exercise 7.38: Find a statement logically equivalent to the negation of the given statement such that negation symbols occur to the right of all quantifiers.

1. $\forall x(x = x)$ _____

2. $\exists x(x < x)$ _____

3. $\forall x\forall y(x = y \wedge x \in y)$ _____

4. $\forall x\exists y(x \leq x \rightarrow y = x)$ _____

5. $\exists x\forall y\exists z\big((x \in y \vee y \in z) \rightarrow x \in z\big)$ _____

Set-theoretic Formulas

An **atomic formula** is an expression of the form $x = y$ or $x \in y$. The letters x and y in these expressions are called **variables**. If we replace the variables in these expressions by specific sets, then we get expressions that are either true or false.

Note: (1) From now on, we will be working with the **language of set theory**. In this language there are just two binary relation symbols and nothing else. The two binary relation symbols are $=$ and $\in$.

(2) In general, languages can have other relation symbols as well as function symbols. For example, in part 3 of Example 7.35 above, we were working in a language with a binary function symbol, $+$, and binary relation symbols, $=$ and $<$.

In this language, there are a lot more types of atomic formulas. Some examples of atomic formulas in this language are $x = y, x < y, x + y = z, x + y < z, (x + y) + z = x + (y + z),...$

(3) When working in the language of set theory, all objects are sets. Therefore, we drop the convention of using capital letters for sets and lowercase letters for elements of sets. In the expression $x \in y$, x and y are both sets.

Example 7.39:

1. The expression $\emptyset = \emptyset$ is true.

2. The expression $\emptyset \in \emptyset$ is false.

3. The expression $\emptyset = \{\emptyset\}$ is false.

4. The expression $\emptyset \in \{\emptyset\}$ is true.

Exercise 7.40: Determine if each of the following expressions is true or false.

1. $\{\emptyset\} = \{\emptyset, \emptyset\}$ _____

2. $\{\emptyset\} \in \{\{\emptyset\}\}$ _____

3. $\{\emptyset\} \in \emptyset$ _____

4. $\{\emptyset, \{\emptyset\}\} \in \{\emptyset, \{\emptyset\}, \{\{\emptyset\}\}, \{\emptyset, \{\emptyset\}\}\}$ _____

To define a **formula** (often called a **first-order formula**), we will need some additional symbols. Specifically, we will need left and right parentheses, symbols for the logical connectives "and," "or," "if...then," "if and only if," and "not," and symbols for the quantifiers "for all" and "there exists." We let S be the set consisting of all the symbols that can appear in formulas:

$$S = \{(,), \wedge, \vee, \rightarrow, \leftrightarrow, \neg, \forall, \exists, =, \in, x, y, z, ...\}.$$

We can now build our list of **formulas** (or **first-order formulas**) from the two types of atomic formulas by using logical connectives and quantifiers.

Example 7.41:

1. Every atomic formula is a formula. For example, $x = y$ and $x \in y$ are formulas.

2. $x = y \wedge x \in y$ is a formula. Here we applied the binary connective $\wedge$ to two atomic formulas to generate a new formula.

3. $x = y \leftrightarrow x \in y$ is a formula. Here we applied the binary connective $\leftrightarrow$ to two atomic formulas to generate a new formula.

4. $\neg x \in y$ is a formula. Here we applied the unary connective $\neg$ to one atomic formula to generate a new formula. We will usually abbreviate this formula as $x \notin y$.

5. $\forall x(x \in y)$ is a formula. Here we applied the quantifier $\forall x$ to one atomic formula to generate a new formula.

6. $(x = y \wedge x \in y) \rightarrow z = w$ is a formula. In part 2 above, we saw that $x = y \wedge x \in y$ is a formula. We then applied the binary connective $\rightarrow$ to this formula and an atomic formula to generate a new formula.

Exercise 7.42: Explain why the following expression is a formula. In other words, describe how it is built up from atomic formulas using logical connectives and quantifiers.

$$\forall x \left(\exists z (z \in x) \wedge \neg \forall y \left(y \in x \wedge \neg \forall w (w \in y \wedge w \notin x) \right) \right)$$

_____

_____

_____

_____

_____

Each occurrence of a variable in a formula can be **free** or **bound**. Bound variables are in the **scope** of a quantifier. Therefore, if a formula has no quantifiers, then all occurrences of variables in the formula are free. If a formula has quantifiers, then we will need to check carefully if each occurrence of each variable is free or bound. A formula ϕ is a **sentence** if no variable occurs free in ϕ.

Example 7.43:

1. In the formula $\forall y (y \in x)$, x is free and y is bound (because y is in the scope of $\forall y$).

2. In the formula $\exists x \forall y (y \in x)$, both x and y are bound (because x is in the scope of $\exists x$ and y is in the scope of $\forall y$). There are no free variables. Therefore, **this formula is a sentence**.

3. In the formula $\forall y (y \in x) \wedge \exists x (x \in y)$, the first occurrence of x is free and the second occurrence of x is bound. Also, the first occurrence of y is bound and the second occurrence of y is free.

4. The formula given in Exercise 7.42 above is a sentence.

Exercise 7.44: Determine if each of the following formulas is a sentence.

1. $x = y$ _____

2. $\forall x \exists y (x \in y)$ _____

3. $\forall x (x \in y) \rightarrow \exists y (y \in x)$ _____

4. $\exists y (\forall x (x \in y) \rightarrow \exists y (y \in x))$ _____

5. $\exists y (\forall x (x \in y) \rightarrow \forall x \exists y (y \in x))$ _____

6. $\exists y (\forall x (x \in y) \rightarrow \forall x \exists y (y \in x)) \vee \forall z (z \in x)$ _____

The Axioms of ZFC

An **axiom** is simply a statement that is assumed to be true. Set theorists, logicians, and philosophers have spent a lot of time and energy trying to decide upon the "correct" collection of axioms. However, deciding whether a statement should be true is a philosophical question and not a mathematical one.

More than one axiomatic system has been developed over the years, but we will stick with one of the more popular and basic ones. Unless otherwise specified, we will be working within the axiomatic system ZFC (Zermelo-Fraenkel set theory **with** Choice). In fact, we have essentially been assuming these axioms throughout this book.

Let's start by writing down each of the axioms of ZF (Zermelo-Fraenkel set theory **without** Choice). We will describe each axiom informally, and then write down the appropriate set-theoretic formula.

Axiom 0 (Empty set): This axiom says that there is a set with no elements. We call this set the empty set, and we use the symbol $\emptyset$ to represent it. Some authors use the symbol { } instead. Symbolically, the axiom looks like this:

$$\exists x \forall y (y \notin x)$$

This can be read as "There is a set x such that for all sets y, y is not in x." In other words, "There is a set x that has no sets in it," or equivalently, "There is a set with no elements."

Set theorists identify the empty set with the natural number 0, so that $0 = \emptyset$.

Axiom 1 (Extensionality): This axiom says that two sets are equal if and only if they have the same elements. Symbolically, the axiom looks like this:

$$\forall x \forall y \big(x = y \leftrightarrow \forall z (z \in x \leftrightarrow z \in y) \big)$$

Note that the Axiom of Extensionality implies that the empty set is unique. To see this, let x and y both be empty sets. Then $\forall z (z \in x \leftrightarrow z \in y)$ is vacuously true. Therefore, by the Axiom of Extensionality, $x = y$.

Axiom 2 (Pairing): This axiom says that given two sets x and y, there is a set whose only elements are x and y. We write this set as $\{x, y\}$. If $x \neq y$, then $\{x, y\}$ has exactly two elements. If $x = y$, then $\{x, y\} = \{x, x\} = \{x\}$, a set with just one element. Symbolically, the axiom looks like this:

$$\forall x \forall y \exists z \forall w \big(w \in z \leftrightarrow (w = x \lor w = y) \big)$$

Example 7.45:

1. If we let $x = 0$ and $y = 0$ (where $0 = \emptyset$), then the pairing axiom gives us the set $\{0, 0\} = \{0\}$. Set theorists identify this set with the natural number 1, so that $1 = \{\emptyset\} = \{0\}$.

2. We can now let $x = 0$ and $y = 1$. The pairing axiom then gives us the set $\{0, 1\} = \{\emptyset, \{\emptyset\}\}$. Set theorists identify this set with the natural number 2. So, $2 = \{\emptyset, \{\emptyset\}\} = \{0, 1\}$ is a set.

3. Here is an example that doesn't have a special identification. Since 1 and 2 are sets, we can use the pairing axiom to form the set $\{1, 2\}$. In its unabbreviated form, $\{1, 2\} = \big\{\{\emptyset\}, \{\emptyset, \{\emptyset\}\}\big\}$.

Exercise 7.46: Let x and y be sets. Use the pairing axiom to explain why the ordered pair (x, y) is a set.

_____

_____

_____

Axiom 3 (Union): This axiom says that given any set x, there is a set whose only elements are the elements of elements of x. We write this set as $\cup x$. If x consists of just two sets a and b, then we write $\cup x$ as $a \cup b$. Symbolically, the axiom looks like this:

$$\forall x \exists y \forall z (z \in y \leftrightarrow \exists w (w \in x \wedge z \in w))$$

Example 7.47:

1. If we let $x = \{2, \{2\}\}$ (this is a set by part 2 of Example 7.45 and the Pairing Axiom), the union axiom gives us the set

    $$\cup x = \cup \{2, \{2\}\} = 2 \cup \{2\} = \{\emptyset, \{\emptyset\}\} \cup \{\{\emptyset, \{\emptyset\}\}\} = \{\emptyset, \{\emptyset\}, \{\emptyset, \{\emptyset\}\}\} = \{0, 1, 2\}.$$

    Set theorists identify this set with the natural number 3. So, we have shown that $3 = \{\emptyset, \{\emptyset\}, \{\emptyset, \{\emptyset\}\}\} = \{0, 1, 2\}$ is a set.

2. If we let $x = \{3, \{3\}\}$ (this is a set by the pairing axiom), the union axiom gives us the set

    $$\cup x = \cup \{3, \{3\}\} = 3 \cup \{3\} = \{\emptyset, \{\emptyset\}, \{\emptyset, \{\emptyset\}\}\} \cup \{\{\emptyset, \{\emptyset\}, \{\emptyset, \{\emptyset\}\}\}\}$$

    $$= \{\emptyset, \{\emptyset\}, \{\emptyset, \{\emptyset\}\}, \{\emptyset, \{\emptyset\}, \{\emptyset, \{\emptyset\}\}\}\} = \{0, 1, 2, 3\} = 4.$$

    This shows that $4 = \{\emptyset, \{\emptyset\}, \{\emptyset, \{\emptyset\}\}, \{\emptyset, \{\emptyset\}, \{\emptyset, \{\emptyset\}\}\}\} = \{0, 1, 2, 3\}$ is a set.

Exercise 7.48:

1. Use the fact that $4 = \{0, 1, 2, 3\}$ is a set to show that $5 = \{0, 1, 2, 3, 4\}$ is a set.

_____

_____

_____

2. Assuming that $n = \{0, 1, \ldots, n-1\}$ is a set, show that $n + 1 = \{0, 1, \ldots, n-1, n\}$ is a set.

_____

_____

_____

Axiom 4 (Power Set): This axiom says that given any set x, there is a set whose only elements are the subsets of x. We write this set as $\mathcal{P}(x)$. Symbolically, the axiom looks like this:

$$\forall x \exists y \forall z (z \in y \leftrightarrow z \subseteq x)$$

Recall that $z \subseteq x$ is an abbreviation for $\forall w(w \in z \to w \in x)$. So, in its unabbreviated form, the power set axiom looks like this:

$$\forall x \exists y \forall z\big(z \in y \leftrightarrow \forall w(w \in z \to w \in x)\big)$$

Example 7.49: Since $2 = \{\emptyset, \{\emptyset\}\}$ is a set, by the power set axiom, we get the set

$$\mathcal{P}(2) = \mathcal{P}(\{\emptyset, \{\emptyset\}\}) = \big\{\emptyset, \{\emptyset\}, \{\{\emptyset\}\}, \{\emptyset, \{\emptyset\}\}\big\}$$

Observe that $\mathcal{P}(2)$ contains the natural numbers 0, 1, and 2, but it also contains the set $\{\{\emptyset\}\}$, which is not equal to a natural number. It follows that $\mathcal{P}(2)$ is **not** a natural number.

Exercise 7.50: Show that $\{0, 1, 2, 3, \{1\}, \{2\}, \{0, 2\}, \{1, 2\}\}$ is a set.

_____

_____

Axiom 5 (Infinity): This axiom says that there is a set x such that $\emptyset \in x$ and whenever $y \in x$, the **successor** of y is also in x, where we define the successor of y to be $S(y) = y \cup \{y\}$. Symbolically, in an abbreviated, easy to read form, the axiom looks like this:

$$\exists x\big(\emptyset \in x \wedge \forall y(y \in x \to y \cup \{y\} \in x)\big)$$

The axiom of infinity provides us with a set that contains all the natural numbers. We can see this from the definition. We are told that $0 = \emptyset \in x$. It then follows that $S(0) = 0 \cup \{0\} = \{0\} = 1 \in x$. And then we have $S(1) = 1 \cup \{1\} = \{0, 1\} = 2 \in x$, and so on.

Exercise 7.51: Write the axiom of infinity in its fully unabbreviated form.

_____

_____

Notice that the Axiom of Infinity provides us only with a set containing the set of natural numbers. What about the set of natural numbers itself? How can we show that it exists? One way to do this is to use the axiom of infinity together with one instance of the next axiom.

Axiom Schema 6 (Bounded Comprehension Schema): Given any set-theoretic formula $\phi(x)$, it would seem natural to have as an axiom that $\{x \mid \phi(x)\}$ is a set. Unfortunately, even for very simple formulas we run into serious problems. For example, if we let $\phi(x)$ be the formula $x \notin x$, we can try to form the "set" $r = \{x \mid x \notin x\}$. Since we are claiming that r is a set, we must have either $r \in r$ or $r \notin r$, but not both. However, by the definition of r, we have $r \in r$ if and only if $r \notin r$. This contradiction is known as **Russell's Paradox**. In light of this contradiction, we must accept the fact that given a formula $\phi(x)$, we cannot insist that the collection $\{x \mid \phi(x)\}$ forms a set. We call the expression $\{x \mid \phi(x)\}$ a **class**. A class that does not form a set is called a **proper class**. For example, $r = \{x \mid x \notin x\}$ is a proper class, which we call the **Russell class**.

We can avoid contradictions like Russell's Paradox by always insisting that new sets we create using formulas always lie inside objects that are already known to be sets. This is called **Bounded Comprehension**. Specifically, if a is already known to be a set and ϕ is any set-theoretic formula, then the bounded comprehension schema tells us that $\{x \in a \mid \phi(x)\}$ is a set. The set a is the **bounding set**.

In its most general form, Bounded Comprehension says the following:

Given sets $b_1, b_2, \ldots, b_n, a$ and a first-order formula ϕ with free variables $w, y_1, y_2, \ldots, y_n, x$, there is a set consisting of all the elements of a that satisfy the formula $\phi(w, y_1, y_2, \ldots, y_n, x)$ when $y_1, y_2, \ldots, y_n, x$ are replaced by $b_1, b_2, \ldots, b_n, a$, respectively. We write this set as $\{w \in a \mid \phi(w, b_1, b_2, \ldots, b_n, a)\}$. The sets $b_1, b_2, \ldots, b_n, a$ are often called **parameters**. Symbolically, given ϕ and parameters $b_1, b_2, \ldots, b_n, a$, the axiom looks like this:

$$\forall x \forall y_1 \forall y_2 \cdots \forall y_n \exists z \forall w \left(w \in z \leftrightarrow \left(w \in x \wedge \phi(w, y_1, y_2, \ldots, y_n, x) \right) \right)$$

Note: The word "schema" indicates that bounded comprehension is not just a single axiom. It is an infinite collection of axioms.

Example 7.52:

1. Let $b_1 = 0 = \emptyset$, $b_2 = 1 = \{\emptyset\}$, $b_3 = 2 = \{\emptyset, \{\emptyset\}\}$, and $a = 100 = \{0, 1, \ldots, 99\}$. Also, let $\phi(w, y_1, y_2, y_3, a)$ be $(w \neq y_1 \wedge y_2 \notin w) \vee y_3 \in w$. By Bounded Comprehension, we get the set $\{w \in 100 \mid (w \neq 0 \wedge 1 \notin w) \vee 2 \in w\} = \{1\} \cup \{3, 4, \ldots, 99\} = \{1, 3, 4, \ldots, 99\}$.

2. Let's use the Axiom of Infinity together with Bounded Comprehension to form the set of natural numbers. By the axiom of infinity, we can let x be a set containing the natural numbers.

   Let $\phi(n)$ be the formula $\forall m(m \notin n) \vee \exists k \forall m(m \in n \leftrightarrow (m \in k \vee m = k))$. This formula expresses that $n = \emptyset$ or $n = S(k)$ for some k.

   Let $\psi(n)$ be $\forall k \left(k \in n \rightarrow \left(\forall m(m \notin k) \vee \exists m(m \in n \wedge \forall j(j \in k \leftrightarrow (j \in m \vee j = m))) \right) \right)$. This formula expresses that for every $k \in n$, $k = \emptyset$ or $k = S(m)$ for some $m \in n$.

   Let $\tau(n)$ be the formula $\phi(n) \wedge \psi(n)$.

   We now use bounded comprehension to form the set $\omega = \{n \in x \mid \tau(n)\}$. I leave it as an optional exercise to show that under the axioms of ZFC this set is the set of natural numbers.

   The symbol "ω" represents a Greek letter that is pronounced "omega." It is the standard symbol that set theorists use for the set of natural numbers. In other words, $\omega = \mathbb{N}$.

3. Given sets a and b, we can define the set $a \cap b$ as $\{w \in a \mid w \in b\}$. Here, we have used the formula $\phi(w, y, x)$ defined by $w \in y$, together with the parameters a and b. The "bounding set" is a. We know this is a set by the Bounded Comprehension Schema.

Exercise 7.53: Let b be a nonempty set. Use the Union Axiom and the Bounded Comprehension Schema to show that $\bigcap b$ is a set.

_____

_____

The next two axioms are a bit technical in nature and more difficult to understand than the previous axioms. Readers can feel free to skip ahead to the Axiom of Choice on the next page. Axiom Schema 7 and Axiom 8 will not be mentioned in the rest of this book. I am including these axioms only for completeness.

Axiom Schema 7 (Replacement Schema): This is a collection of axioms, which says that the image of a set under a function is also a set. A formula $\phi(x, z)$ represents a function if

$$\forall x \forall w \forall v \left((\phi(x, w) \land \phi(x, v)) \to w = v \right).$$

The Replacement Schema will allow parameters (just like the Bounded Comprehension Schema). Therefore, the expression that ϕ represents a function should technically be written as follows:

$$\forall x \forall w \forall v \forall y_1 \forall y_2 \cdots \forall y_n \left((\phi(x, w, y_1, y_2, \ldots, y_n) \land \phi(x, v, y_1, y_2, \ldots, y_n)) \to w = v \right)$$

We will abbreviate this formula by simply writing "$\phi(x, z)$ is a function." Note that there may be other variables in the formula even though we are not mentioning them explicitly.

Informally, the Replacement Schema will then say "If $\phi(x, z)$ is a function, then for any given set A acting as the domain of the function, the image of A under the function is a set." We can write this more formally as follows:

$$\phi(x, z) \text{ is a function} \to \forall A \exists B \forall s (s \in B \leftrightarrow \exists t (t \in A \land \phi(t, s))$$

Just for fun, let's write the unabbreviated form of this formula. To save space, let's let $\boldsymbol{y}$ represent $y_1, y_2, \ldots, y_n$ and $\forall \boldsymbol{y}$ represent $\forall y_1 \forall y_2 \cdots \forall y_n$. The axiom then looks like this:

$$\forall x \forall w \forall v \forall \boldsymbol{y} \left((\phi(x, w, \boldsymbol{y}) \land \phi(x, v, \boldsymbol{y})) \to w = v \right) \to \forall A \forall \boldsymbol{y} \exists B \forall s (s \in B \leftrightarrow \exists t (t \in A \land \phi(t, s, \boldsymbol{y}))$$

In branches of mathematics outside of set theory, it's hard to find much use for the Replacement Schema. However, for set theorists, it is useful, as it is needed to show that every ordinal is a set. We will learn about the ordinals in Lesson 8. In this book, we will be discussing the ordinals informally, and so, we will not be mentioning the Replacement Schema again.

Axiom 8 (Foundation): This axiom says that every nonempty set has an $\in$-least element. Symbolically, the axiom looks like this:

$$\forall x (\exists z (z \in x) \to \exists z (z \in x \land \forall y (y \in z \to y \notin x))$$

Like the Replacement Schema, the Axiom of Foundation is not used much outside of set theory, and in fact, much of set theory can be developed without it. Some mathematicians actually take the opposite approach and adopt an axiom called "the Axiom of Antifoundation." One interesting consequence of antifoundation is that there exists a unique set x such that $x = \{x\}$. This set is quite strange (and it does not exist under the Axiom of Foundation). Clearly, we have $x \in x$. This leads to an infinite descending $\in$-chain $x \ni x \ni x \ni x \ni \cdots$ (if the backwards "epsilon" confuses you, we can also write this sequence as $\cdots \in x \in x \in x \in x$).

Axioms 0 through 8 form an axiom system known as ZF. This stands for "Zermelo-Fraenkel." Sometimes one may wish to leave out the Axiom of Foundation. In this case, we have the axiom system ZF⁻.

We complete our list of axioms with one that has caused much controversy over the years.

Axiom 9 (Axiom of Choice): One version of this axiom says that for every set X of nonempty pairwise disjoint sets, there is a set Y that contains exactly one element from each set in X. The set Y is sometimes called a **selector** for X.

Let's try to write out the unabbreviated first-order formula. First, let ϕ be the statement "X is a set of nonempty pairwise disjoint sets." Then ϕ can be written as follows:

$$\forall x\big(x \in X \to \exists y(y \in x)\big) \wedge \forall x \forall y\big((x \in X \wedge y \in X \wedge x \neq y) \to \forall z(z \notin x \vee z \notin y)\big).$$

Now, let ψ be the statement "There is a set Y that contains exactly one element from each set in X." Then ψ can be written as follows:

$$\exists Y \forall x\big(x \in X \to \exists! y(y \in x \wedge y \in Y)\big).$$

The exclamation point (!) is an abbreviation used for uniqueness. $\exists! y$ is read "there is a unique y." In general, $\exists! y\big(\tau(y)\big)$ is an abbreviation for $\exists y\big(\tau(y) \wedge \forall z(\tau(z) \to z = y)\big)$. So, $\exists! y(y \in x \wedge y \in Y)$ is an abbreviation for $\exists y\big(y \in x \wedge y \in Y \wedge \forall z((z \in x \wedge z \in Y) \to z = y)\big)$.

Putting this all together, the unabbreviated formula expressing the Axiom of Choice is the following:

$$\forall X \Bigg(\bigg(\forall x\big(x \in X \to \exists y(y \in x)\big) \wedge \forall x \forall y\big((x \in X \wedge y \in X \wedge x \neq y) \to \forall z(z \notin x \vee z \notin y)\big)\bigg) \to$$

$$\exists Y \forall x \bigg(x \in X \to \exists y \big(y \in x \wedge y \in Y \wedge \forall z((z \in x \wedge z \in Y) \to z = y)\big)\bigg)\Bigg).$$

The Axiom of Choice can be phrased as follows: "For every set X of nonempty pairwise disjoint sets, there is a set Y such that for each $x \in X$, there is a unique $y_x \in x$ such that $y_x \in Y$. Notice that $Y \subseteq \bigcup X$. So, we can define a function $f: X \to \bigcup X$ by letting $f(x) = y_x$. This function is called a **choice function**. It chooses one element from each set in X.

In fact, the Axiom of Choice is equivalent to the following statement: "For every set X of nonempty sets, there is a choice function $f: X \to \bigcup X$" (notice the absence of the phrase "pairwise disjoint"). You will be asked to show that this statement is equivalent to AC in Problem 85 below.

Example 7.54:

1. Let $X = \big\{\{0, 1, 2\}, \{a, b, c, d\}, \{\Delta, \square\}\big\}$ and let $f = \{(\{0, 1, 2\}, 1), (\{a, b, c, d\}, d), (\{\Delta, \square\}, \Delta)\}$. We see that $f: X \to \bigcup X$ satisfies $f(\{0, 1, 2\}) = 1$, $f(\{a, b, c, d\}) = d$, and $f(\{\Delta, \square\}) = \Delta$. This function f is a choice function. The corresponding selector Y is $\{1, d, \Delta\}$. Note that the Axiom of Choice is **not** needed to provide a choice function for this example. In fact, whenever X is a finite set of sets, we can always find a choice function without resorting to the axiom of choice.

2. Let $X = \mathcal{P}(\mathbb{N}) \setminus \emptyset$ and define $f: X \to \mathbb{N}$ by $f(S) = $ the least element of S. Then f is a choice function. For example, if $\mathbb{O}$ is the set of odd natural numbers, then $f(\mathbb{O}) = 1$. If $\mathbb{P}$ is the set of primes, then $f(\mathbb{P}) = 2$. The axiom of choice is actually **not** needed for this example. In this case, we could define a choice function by using the fact that every subset of natural numbers has a least element (this is known as the **Well Ordering Principle** of $\mathbb{N}$).

133

More generally, if a set A can be ordered in such a way that every subset of S has a least element, then the axiom of choice will not be needed to define a choice function on $\mathcal{P}(A) \setminus \emptyset$. Once again, we can simply define $f : \mathcal{P}(A) \setminus \emptyset \to A$ by $f(S) =$ the least element of S.

3. For each natural number n, let A_n be a set with two distinct elements. Can we *explicitly define* a choice function $f : X \to \bigcup X$? The answer is no. Even though all the sets in X are finite, we do not have enough information about the sets to explicitly describe a choice function. However, if we assume the axiom of choice, then we can say that a choice function exists. Furthermore, we can feel free to use this choice function to prove other results. This example gives a little insight into why the axiom of choice was so controversial when it was first introduced. It's like magic. All of a sudden, we have this function that we can't describe. We weren't able to come up with the function ourselves, so we simply said, "No big deal—the axiom of choice will take care of this for us. Here's your choice function." Someone might ask, "Well, what does it look like?" And you would have to respond, "I have no idea, but here it is."

   We can replace the natural numbers with any set here. For example, we can let A_r be a set with two distinct elements for each $r \in \mathbb{R}$. Even though $\mathbb{R}$ is uncountable, the axiom of choice still gives us a choice function $f : X \to \bigcup X$. The axiom of choice is indeed very powerful.

Axioms 0 through 9 form the axiom system known as ZFC. This stands for "Zermelo-Fraenkel with Choice."

LEVEL 1

Determine whether each of the following sentences is an atomic statement, a compound statement, or not a statement at all:

1. Henry does not know where he is going.

2. What are you doing?

3. Stop doing that!

4. $x \neq 53$.

5. I like the song *Crimson and Clover*.

6. If lobsters are birds, then we have a problem.

7. $5 = 6$ or $5 \neq 6$.

8. This sentence has five words.

9. A quadrilateral is a rhombus if and only if the diagonals of the quadrilateral are perpendicular.

10. I cannot speak Japanese, but I can speak Italian.

What is the negation of each of the following statements?

11. A tomato is a fruit.

12. Cats have nine lives.

13. $76 \geq 72$.

14. You will not believe what I just heard.

15. The function f has a discontinuity at $x = 3$.

16. The fundamental group of the circle is isomorphic to the set of integers.

17. Some birds can fly.

18. Every summer night is warm.

Let p represent the statement "5 is an odd integer," let q represent the statement "Brazil is in Europe," and let r represent the statement "A lobster is an insect." Rewrite each of the following symbolic statements in words, and state the truth value of each statement:

19. $p \lor q$

20. $\neg r$

21. $p \to q$

22. $p \leftrightarrow r$

23. $\neg q \land r$

24. $\neg(p \land q)$

25. $\neg p \lor \neg q$

26. $(p \land q) \to r$

Consider the compound sentence "You can have a cookie or ice cream." In English this would most likely mean that you can have one or the other but not both. The word "or" used here is generally called an "exclusive or" because it excludes the possibility of both. The disjunction is an "inclusive or."

27. Using the symbol $\oplus$ for exclusive or, draw the truth table for this connective.

28. Using only the logical connectives $\neg$, $\land$, and $\lor$, produce a statement using the propositional variables p and q that has the same truth values as $p \oplus q$.

Determine whether each occurrence of each variable in the given formula is free or bound. Is the given formula a sentence?

29. $x \in y$

30. $\forall x(x \in y)$

31. $\exists y\big(x = y \to \forall x(y \in x)\big)$

32. $\forall x \forall y \exists z\big((x \in y \land x \in z) \leftrightarrow x = z\big) \lor \big(x \neq y \to \forall z(x \in z)\big)$

Consider the four distinct propositional variables p, q, r, and s.

33. How many different truth assignments are there for this list of propositional variables?

34. How many different truth assignments are there for this list of propositional variables such that p is true and q is false?

35. How many different truth assignments are there for this list of propositional variables such that q, r, and s are all true?

36. How many different truth assignments are there for a list of 5 propositional variables?

Let p, q, and r represent true statements. Compute the truth value of each of the following compound statements:

37. $(p \lor q) \lor r$

38. $(p \lor q) \land \neg r$

39. $\neg p \rightarrow (q \lor r)$

40. $\neg(p \leftrightarrow \neg q) \land r$

41. $\neg[p \land (\neg q \rightarrow r)]$

42. $\neg[(\neg p \lor \neg q) \leftrightarrow \neg r]$

43. $p \rightarrow (q \rightarrow \neg r)$

44. $\neg[\neg p \rightarrow (q \rightarrow \neg r)]$

Determine if each of the following statements is a tautology, a contradiction, or neither.

45. $p \land p$

46. $p \land \neg p$

47. $(p \lor \neg p) \rightarrow (p \land \neg p)$

48. $\neg(p \lor q) \leftrightarrow (\neg p \land \neg q)$

49. $p \rightarrow (\neg q \land r)$

50. $(p \leftrightarrow q) \rightarrow (p \rightarrow q)$

Assume that the given compound statement is true. Determine the truth value of each propositional variable.

51. $p \wedge q$

52. $\neg(p \rightarrow q)$

53. $p \leftrightarrow [\neg(p \wedge q)]$

54. $[p \wedge (q \vee r)] \wedge \neg r$

Let p represent a true statement. Decide if this is enough information to determine the truth value of each of the following statements. If so, state that truth value.

55. $p \vee q$

56. $p \rightarrow q$

57. $\neg p \rightarrow \neg(q \vee \neg r)$

58. $\neg(\neg p \wedge q) \leftrightarrow p$

59. $(p \leftrightarrow q) \leftrightarrow \neg p$

60. $\neg[(\neg p \wedge \neg q) \leftrightarrow \neg r]$

61. $[(p \wedge \neg p) \rightarrow p] \wedge (p \vee \neg p)$

62. $r \rightarrow [\neg q \rightarrow (\neg p \rightarrow \neg r)]$

For each of the following pairs of statements A and B, show that $A \equiv B$.

63. $A = p \wedge q, B = q \wedge p$

64. $A = (p \vee q) \vee r, B = p \vee (q \vee r)$

65. $A = p \wedge (q \vee r), B = (p \wedge q) \vee (p \wedge r)$

66. $A = (p \vee q) \wedge p, B = p$

67. $A = p \leftrightarrow q, B = (p \rightarrow q) \wedge (q \rightarrow p)$

68. $A = \neg(p \rightarrow q), B = p \wedge \neg q$

Simplify each statement.

69. $p \lor (p \land \neg p)$

70. $(p \land q) \lor \neg p$

71. $\neg p \to (\neg q \to p)$

72. $(p \land \neg q) \lor p$

73. $[(q \land p) \lor q] \land [(q \lor p) \land p]$

LEVEL 5

Without drawing a truth table or using List 7.28, show that each of the following is a tautology.

74. $[p \land (q \lor r)] \leftrightarrow [(p \land q) \lor (p \land r)]$

75. $\big[[(p \land q) \to r] \to s\big] \to [(p \to r) \to s]$

Let n be a positive integer (in other words, n is one of the numbers 1, 2, 3, 4, ...) and let A be a statement involving n propositional variables. Determine how many rows are in the truth table for A if n is equal to each of the following:

76. $n = 6$

77. $n = 10$

78. n is an arbitrary positive integer (provide an explicit expression involving n.)

Use the axioms of ZF to show each of the following:

79. If x, y, and z are sets, then the ordered triple (x, y, z) is a set.

80. If $x_1, x_2, \ldots, x_{n+1}$ are sets and every n-tuple of sets is a set, then $(x_1, x_2, \ldots, x_n, x_{n+1})$ is a set.

81. If A and B are sets, then $\mathcal{P}\big(\mathcal{P}(A \cup B)\big)$ is a set.

82. If A and B are sets, then $A \times B$ is a set.

139

83. Determine a tautology or contradiction containing *at least* three propositional variables and *at least* three logical connectives so that the truth values for all truth assignments can be evaluated with *no more than* 5 computations and such that *at least* 3 computations are required. Write out your compound statement, followed by your 3 to 5 computations.

84. Let T be a truth table. Explain why there is a statement A involving only the logical connectives $\wedge$, $\vee$, and $\neg$ such that the truth table of A is T.

    Hint: For example, let T be the following truth table:

p	q	?
T	T	T
T	F	F
F	T	T
F	F	F

    If we let A be the statement $(p \wedge q) \vee (\neg p \wedge q)$, then the truth table of A is T.

85. Show that the following two versions of AC are equivalent: (i) If X is a set of nonempty sets, then there is a function $f \colon X \to \bigcup X$ such that for each $x \in X$, $f(x) \in x$; (ii) If X is a set of nonempty pairwise disjoint sets, then there is a set Y that contains exactly one element from each set in X.

LESSON 8
ORDINALS AND CARDINALS

Well-ordered Sets

Recall from Lesson 3 that an **ordered set** is a pair $(A, <)$, where A is a set and $<$ is a strict linear ordering on A. In other words, $<$ is a binary relation on A such that

- for all $a, b, c \in A$, $a < b$ and $b < c$ imply $a < c$ (we say that $<$ is **transitive** on A).

- for all $a, b \in A$, exactly one of $a < b$, $b < a$, or $a = b$ holds (we say that $<$ is **trichotomous** on A).

An ordered set $(A, <)$ is called **well-ordered** if for every nonempty subset $B \subseteq A$, there is an element $b \in B$ such that for all $c \in B$ with $c \neq b$, $b < c$. This element $b \in B$ is called the $<$-**least element** of B.

Example 8.1:

1. $(\mathbb{N}, <)$ is well-ordered. This is known as the **Well Ordering Principle** of $\mathbb{N}$. We can visualize this ordering as follows:

$$0 < 1 < 2 < 3 < 4 < 5 < 6 < \cdots$$

   Notice how $\mathbb{N}$ with its usual ordering has a least element 0 (let's call this the zeroth element), a first element 1, a second element 2, a third element 3, and so on. $(\mathbb{N}, <)$ is the most basic example of an infinite well-ordered set.

2. $(\mathbb{Z}, <)$, $(\mathbb{Q}, <)$, and $(\mathbb{R}, <)$ are **not** well-ordered. None of $\mathbb{Z}$, $\mathbb{Q}$, or $\mathbb{R}$ have a $<$-least element.

3. $([0, 1], <)$ is **not** well-ordered. Although $[0, 1]$ has the $<$-least element 0, the subset $(0, 1)$ has no $<$-least element.

   Note that when we write an interval such as $[0, 1]$ without any additional comments, we are usually thinking of this interval as a subset of $\mathbb{R}$ (see Lesson 2). We will write $[0, 1] \cap \mathbb{Q}$ for the corresponding subset of $\mathbb{Q}$. It's worth noting that $([0, 1] \cap \mathbb{Q}, <)$ is not well-ordered either.

4. Define the relation $<'$ (read "less than prime") on $\mathbb{N}$ by

$$<' = \{(m, n) \mid (m \neq 0, n \neq 0, \text{ and } m < n) \vee (m \neq 0 \text{ and } n = 0)\}.$$

   We can visualize this ordering as follows:

$$1 <' 2 <' 3 <' 4 <' 5 <' \cdots <' 0$$

   In words, (i) all nonzero natural numbers are placed in their usual order and (ii) all nonzero natural numbers are defined to be less than 0. Instead of 0 being the least element in this ordering, it is the greatest element.

   Let's check that this relation is transitive. Suppose that $m <' n$ and $n <' k$. If m, n, and k are all nonzero, then $<'$ is the same as $<$, and we can use the transitivity of $<$ to see that $m <' k$. The only other possible case is that m, n are nonzero and $k = 0$. In this case, $m <' k$ by the definition of $<'$ (every nonzero natural number is $<'$ 0).

Let's next check that $<'$ is trichotomous. Let $m, n \in \mathbb{N}$. Since $<'$ agrees with $<$ when $m, n \neq 0$, we may assume that one of them is 0, say $m = 0$. By the definition of $<'$, either $n = 0$ or $n <' m$ (but not both), completing the argument.

So, we have shown that $(\mathbb{N}, <')$ is an ordered set.

In fact, $(\mathbb{N}, <')$ is well-ordered. To see this, let B be a nonempty subset of $\mathbb{N}$. If $B = \{0\}$, then 0 is the $<'$-least element of B. Otherwise, the $<$-least element of $B \setminus \{0\}$ is the $<'$-least element of B.

Exercise 8.2: Let $A = \mathbb{N} \cup \{\star\}$, where $\star$ is an element that is not equal to any natural number and define $<^\star$ on A by

$$<^\star = \{(m, n) \mid (m, n \in \mathbb{N} \text{ and } m < n) \vee (m \in \mathbb{N} \text{ and } n = \star)\}.$$

1. Is $<^\star$ transitive on A?   _____

2. Is $<^\star$ trichotomous on A?   _____

3. Is $(A, <^\star)$ an ordered set?   _____

4. Is $(A, <^\star)$ well-ordered?   _____

5. Draw a visualization of this ordering as we did in parts 1 and 4 of Example 8.1.

Well-ordered sets are **isomorphic** if there is an "order preserving" bijection between them.

More formally, well-ordered sets $(A, <_A)$ and $(B, <_B)$ are isomorphic, written $(A, <_A) \cong (B, <_B)$ (often abbreviated as $A \cong B$), if there is a bijection $f: A \to B$ such that for all $x, y \in A$, we have

$$x <_A y \text{ if and only if } f(x) <_B f(y).$$

The order preserving bijection f defined above is called an **isomorphism** from A to B.

Mathematicians like to identify well-ordered sets that are isomorphic and essentially treat them as if they were identical. After all, they do look exactly the same – only the names have been changed.

Example 8.3:

1. $(\{0, 1, 2\}, <)$, where $<$ is the usual ordering, is well-ordered. We can visualize this ordering as follows:

$$0 < 1 < 2$$

Let's also define an ordering $\prec$ on the set $\{a, b, c\}$ in the most natural way (alphabetically). Then $(\{a, b, c\}, \prec)$ is also well-ordered. We can visualize this ordering as follows:

$$a \prec b \prec c$$

The well-ordered sets $(\{0, 1, 2\}, <)$ and $(\{a, b, c\}, \prec)$ are isomorphic.

Symbolically, we can write $(\{0,1,2\},<) \cong (\{a,b,c\},\prec)$ or $\{0,1,2\} \cong \{a,b,c\}$. There is exactly one isomorphism f from $\{0,1,2\}$ to $\{a,b,c\}$, namely $f = \{(0,a),(1,b),(2,c)\}$. It can be visualized as follows:

$$0 < 1 < 2$$
$$\downarrow \quad \downarrow \quad \downarrow \quad f$$
$$a \prec b \prec c$$

In fact, all well-ordered sets with three elements are isomorphic. So, there is essentially just one well-ordered set with three elements. Let's take a moment to look at one more way to describe this well-ordered set. The following description is preferred by set theorists. The reason for this will be explored a bit later in this lesson.

If we define $0 = \emptyset$, $1 = \{\emptyset\} = \{0\}$, $2 = \{\emptyset, \{\emptyset\}\} = \{0,1\}$, and $3 = \{\emptyset, \{\emptyset\}, \{\emptyset, \{\emptyset\}\}\} = \{0,1,2\}$, then $(3, \in)$ is another well-ordered set with three elements. Here are a few ways to visualize this well-ordered set:

$$\emptyset \in \{\emptyset\} \in \{\emptyset, \{\emptyset\}\}$$
$$0 \in \{0\} \in \{0,1\}$$
$$0 \in 1 \in 2$$
$$0 <_{\mathbb{N}} 1 <_{\mathbb{N}} 2$$
$$0 < 1 < 2$$

See the end of Lesson 3 for more details on this ordering.

The well-ordered set $(3, \in)$ is an example of an **ordinal**. Specifically, $(3, \in)$ is the unique ordinal with three elements. Set theorists usually abbreviate the ordinal $(3 \in)$ with the natural number 3. Ordinals are the "numbers" that set theorists have chosen to represent the well-ordered sets. We will see a precise definition of ordinal shortly.

2. For each natural number n, there is essentially exactly one well-ordered set of size n. The well-ordered set of size zero is $0 = \emptyset$ with the empty well-ordering. The well-ordered set of size one is $1 = \{\emptyset\} = \{0\}$. The well-ordered set of size two is $2 = \{\emptyset, \{\emptyset\}\} = \{0,1\}$. As we saw in part 1 above, the well-ordered set of size three is $3 = \{\emptyset, \{\emptyset\}, \{\emptyset, \{\emptyset\}\}\} = \{0,1,2\}$. For any natural number n, the well-ordered set of size $n + 1$ is $n + 1 = \{0,1,2,\ldots,n\} = n \cup \{n\}$. Each of these sets is an ordinal. They are all well-ordered by the membership relation $\in$. Every finite well-ordered set is isomorphic to one of these finite ordinals. Specifically, if $(A, <_A)$ is a well-ordered set such that $|A| = n$, then $(A, <_A) \cong n$, where n is an abbreviation for $(n, \in)$. Here are a few ways to visualize the well-ordered set $n + 1$:

$$0 \in \{0\} \in \{0,1\} \in \{0,1,2\} \in \cdots \in \{0,1,2,\ldots,n-1\}$$
$$0 \in 1 \in 2 \in 3 \in \cdots \in n$$
$$0 <_{\mathbb{N}} 1 <_{\mathbb{N}} 2 <_{\mathbb{N}} 3 <_{\mathbb{N}} \cdots <_{\mathbb{N}} n$$
$$0 < 1 < 2 < 3 < \cdots < n$$

Once again, see the end of Lesson 3 for more details on this ordering.

3. Set theorists use the greek letter ω (pronounced "omega") when we are thinking of the set of natural numbers $\omega = \{0, 1, 2, 3, \dots\}$ using its set theoretic definition (see part 2 above or the end of Lesson 3). Let's show that $\omega = \{0, 1, 2, 3, \dots\}$ is well-ordered by $\in$.

To see that $\in$ is transitive on ω, let $j, k, l \in \omega$, and choose $m \in \omega$ with $j, k, l \in m$. If $j \in k$ and $k \in l$, then by transitivity of $\in$ on m, we have $j \in l$.

To see that $\in$ is trichotomous on ω, let $j, k \in \omega$, and choose $m \in \omega$ with $j, k \in m$. Since $\in$ is trichotomous on m, either $j \in k$, $j = k$, or $k \in j$ and exactly one of these holds.

Finally, let A be a nonempty subset of ω, and choose $n \in A$. If $n \cap A = \emptyset$, then n is the $\in$-least element of A (If $m \in A$ with $m \in n$, then $m \in n \cap A$, contradicting $n \cap A = \emptyset$). Otherwise, $n \cap A$ is a nonempty subset of n, and since n is well-ordered by $\in$, $n \cap A$ has an $\in$-least element m. Then m is the $\in$-least element of A (If $k \in A$ with $k \in m$, then $k \in n$ because n is transitive, and so, m is not the $\in$-least element of $n \cap A$).

4. Define the relation $<_*$ on $\mathbb{N}$ by

$$<_* = \{(m, n) \mid (m \text{ and } n \text{ have the same parity and } m < n) \vee (m \text{ is even and } n \text{ is odd})\}$$

We can visualize this ordering as follows:

$$0 <_* 2 <_* 4 <_* 6 <_* 8 <_* \cdots <_* 1 <_* 3 <_* 5 <_* \cdots$$

In words, we are insisting that all even natural numbers are less than all odd natural numbers, while the evens and odds themselves are placed in their usual order within themselves.

$(\mathbb{N}, <_*)$ is a well-ordered set that is **not** isomorphic to $(\mathbb{N}, <)$ (where $<$ is the usual order on $\mathbb{N}$). To see this, suppose that $f \colon \mathbb{N} \to \mathbb{N}$ satisfies $n < m$ if and only if $f(n) <_* f(m)$. Suppose also that $1 \in \operatorname{ran} f$. Let $a \in \mathbb{N}$ satisfy $f(a) = 1$. If $a = 0$, then $\operatorname{ran} f$ contains no even integers, and so, f is not surjective. If $a \neq 0$, let $f(a-1) = b$. Since $a - 1 < a$, we have $b <_* 1$. So, b is even. Therefore, $b + 2$ is also even. So, $b <_* b + 2 <_* 1$. It follows that $b + 2 \notin \operatorname{ran} f$. So, once again, f is not surjective. Therefore, f is not an isomorphism, and so, $(\mathbb{N}, <)$ is **not** isomorphic to $(\mathbb{N}, <_*)$.

Exercise 8.4: Determine if each of the following pairs of well-ordered sets are isomorphic. If so, describe an isomorphism between them. If not, explain why.

1. $(4, \in)$ and $(\{a, b, c, d\}, <)$, where $<$ orders $\{a, b, c, d\}$ alphabetically

_____

_____

_____

2. $(5, \in)$ and $(6, \in)$

_____

_____

_____

3. $(\mathbb{N}, <')$ (from part 4 of Example 8.1) and $(A, <^*)$ from Exercise 8.2

_____

_____

_____

4. $(500, \in)$ and $(\omega, \in)$

_____

_____

_____

5. $(\omega, \in)$ and $(\mathbb{N}, <')$ (from part 4 of Example 8.1)

_____

_____

_____

The collection of all well-ordered sets has a particularly nice structure. As we will soon see, any two well-ordered sets are either isomorphic or one is isomorphic to an "initial segment" of the other.

Let $(A, <)$ be a well-ordered set with least element a. If $b \in A$ with $b \neq a$, then the set of **predecessors** of b is $\text{pred}(A, b) = [a, b) = \{x \in A \mid x < b\}$. The set of predecessors of a is $\emptyset$. That is, $\text{pred}(A, a) = [a, a) = \emptyset$. A subset $S \subseteq A$ is an **initial segment** of A if for all $b \in S$, $[a, b) \subseteq S$ (or equivalently, $\text{pred}(A, b) \subseteq S$).

Example 8.5:

1. Consider $\omega = \{0, 1, 2, \dots\}$ ordered by the $\in$ relation. The set of predecessors of 0 is $\emptyset$. The set of predecessors of 1 is $[0, 1) = \{0\}$. The set of predecessors of 2 is $[0, 2) = \{0, 1\}$. In general, the set of predecessors of n is $[0, n) = \{0, 1, 2, \dots, n - 1\}$. Each $n \in \omega$ is an initial segment of ω. The set $A = \{0, 1, 3, 4, 5, 6\}$ is **not** an initial segment of ω because $3 \in A$, but the set of predecessors of 3 is $[0, 3) = \{0, 1, 2\}$, which is not a subset of A because $2 \in [0, 3)$, but $2 \notin A$.

2. Consider the well-ordered set $(\mathbb{N}, <')$ from Part 4 of Example 8.1. Recall that the well ordering $<'$ is defined on $\mathbb{N}$ by

$$<' = \{(m, n) \mid (m \neq 0, n \neq 0, \text{and } m < n) \vee (m \neq 0 \text{ and } n = 0)\}.$$

   Once again, we can visualize this ordering as follows:

$$1 <' 2 <' 3 <' 4 <' 5 <' \cdots <' 0$$

   The set of predecessors of 3 is $[1, 3) = \{1, 2\}$.

   The set of predecessors of 0 is $[1, 0) = \{1, 2, 3, 4, \dots\} = \mathbb{N}^*$, the set of nonzero natural numbers.

   Notice that there is an isomorphism f from $(\omega, \in)$ to an initial segment of $(\mathbb{N}, <')$ defined by $f(n) = n + 1$. Observe that $f[\omega] = [1, 0) = \mathbb{N}^*$.

3. Given a well-ordered set $(A, <)$, A is trivially an initial segment of itself.

Exercise 8.6: Consider the well-ordered set $(\mathbb{N}, <_*)$ from Part 4 of Example 8.3. Find each of the following:

1. $[0, 8)$ _____

2. $[0, 5)$ _____

3. $\text{pred}(\mathbb{N}, 0)$ _____

4. $\text{pred}(\mathbb{N}, 14)$ _____

5. $\text{pred}(\mathbb{N}, 15)$ _____

6. an isomorphism f from $(\omega, \in)$ to an initial segment of $(\mathbb{N}, <_*)$

_____

7. $f[\omega]$, where f was defined in part 6 _____

If S is an initial segment of A and $S \neq A$, we call S a **proper initial segment** of A.

> **Well-ordered Set Fact 1:** If S is a proper initial segment of a well-ordered set A, then there is $b \in A$ such that $S = \text{pred}(A, b)$.

> **Well-ordered Set Fact 2:** Two well-ordered sets are either isomorphic or one is isomorphic to a proper initial segment of the other one.

Well-ordered Set Fact 2 essentially says that all well-ordered sets are built up the same way. There's a least element, then a next element, then a third element, and so on... eventually we get to the first "limiting element." This is an element that is greater than the first countably many elements (note that finite well-ordered sets and the well-ordered set $(\omega, \in)$ do not have any "limiting elements"). Let's call this the ωth (pronounced "omegeth") element. We then continue to the "omega plus first" element, ... and so on. The following picture shows what an isomorphism from a well-ordered set $(A, <_A)$ to a proper initial segment of a well-ordered set $(B, <_B)$ must look like.

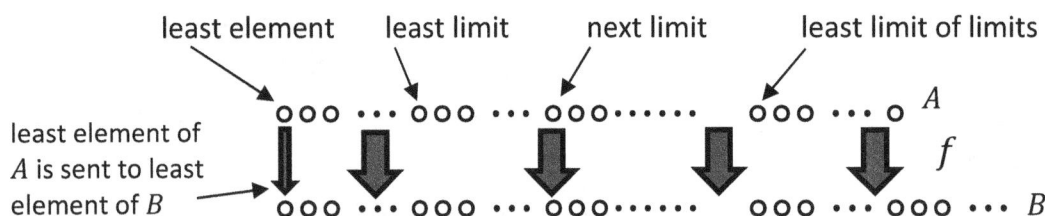

Transitive Sets

A set x is **transitive** if every element of x is also a subset of x. Symbolically, we can write

$$\forall y(y \in x \rightarrow y \subseteq x)$$

Notes: (1) Another way to say that a set x is transitive is to say that every element of an element of x is an element of x. Symbolically, $\forall y \forall z\big((z \in y \land y \in x) \rightarrow z \in x\big)$.

(2) From here on out, all of the objects we will be considering are sets. Therefore, we drop the convention of using capital letters for sets and lowercase letters for elements of sets. In the expression $a \in A$, the symbols a and A will both be representing sets. Furthermore, after looking at an element a in the set A, since a is itself a set, we may want to look at elements of a. So, maybe we have $b \in a$. Since b is also a set, we can look at an element $c \in b$. In this situation, it adds little value to capitalize the name of the first set.

Example 8.7:

1. $0 = \emptyset$ is transitive. Indeed, $\emptyset$ has no elements, and so, it is vacuously true that every element of $\emptyset$ is a subset of $\emptyset$.

2. $1 = \{\emptyset\}$ is transitive. To see this, note that the only element of $\{\emptyset\}$ is $\emptyset$, and $\emptyset$ is a subset of every set. In particular, $\emptyset \subseteq \{\emptyset\}$.

3. $2 = \{\emptyset, \{\emptyset\}\}$ is transitive. The elements of $\{\emptyset, \{\emptyset\}\}$ are $\emptyset$ and $\{\emptyset\}$. As in 2, we observe that $\emptyset$ is a subset of every set, and so, $\emptyset \subseteq \{\emptyset, \{\emptyset\}\}$. The only element of $\{\emptyset\}$ is $\emptyset$, and $\emptyset \in \{\emptyset, \{\emptyset\}\}$. Therefore, $\{\emptyset\} \subseteq \{\emptyset, \{\emptyset\}\}$.

4. $\{\emptyset, \{\{\emptyset\}\}\}$ is **not** transitive. We have $\{\emptyset\} \in \{\{\emptyset\}\} \in \{\emptyset, \{\{\emptyset\}\}\}$, but $\{\emptyset\} \notin \{\emptyset, \{\{\emptyset\}\}\}$. Therefore, $\{\{\emptyset\}\} \in \{\emptyset, \{\{\emptyset\}\}\}$, but $\{\{\emptyset\}\} \not\subseteq \{\emptyset, \{\{\emptyset\}\}\}$.

5. $\mathcal{P}(\{\emptyset, \{\emptyset\}\}) = \{\emptyset, \{\emptyset\}, \{\{\emptyset\}\}, \{\emptyset, \{\emptyset\}\}\}$ is transitive. The elements of $\{\emptyset, \{\emptyset\}, \{\{\emptyset\}\}, \{\emptyset, \{\emptyset\}\}\}$ are $\emptyset$, $\{\emptyset\}$, $\{\{\emptyset\}\}$, and $\{\emptyset, \{\emptyset\}\}$. By the same reasoning as in 3, $\emptyset$ and $\{\emptyset\}$ are subsets of $\{\emptyset, \{\emptyset\}, \{\{\emptyset\}\}, \{\emptyset, \{\emptyset\}\}\}$. The only element of $\{\{\emptyset\}\}$ is $\{\emptyset\}$, which is in $\{\emptyset, \{\emptyset\}, \{\{\emptyset\}\}, \{\emptyset, \{\emptyset\}\}\}$. The set $\{\emptyset, \{\emptyset\}\}$ has two elements: $\emptyset$ and $\{\emptyset\}$, both of which are in $\{\emptyset, \{\emptyset\}, \{\{\emptyset\}\}, \{\emptyset, \{\emptyset\}\}\}$.

6. Every natural number is a transitive set. We saw in 1, 2, and 3 above that $0 = \emptyset$, $1 = \{\emptyset\}$, and $2 = \{\emptyset, \{\emptyset\}\}$ are transitive. Assuming that the natural number k is transitive, let $j \in k + 1$ and $m \in j$. Since $k + 1 = k \cup \{k\}$, we have $j \in k$ or $j \in \{k\}$. If $j \in k$, then we have $m \in j \in k$. Since k is transitive, $m \in k$. So, $m \in k \cup \{k\} = k + 1$. If $j \in \{k\}$, then $j = k$. So, $m \in k$, and again, we have $m \in k \cup \{k\} = k + 1$.

7. The set of natural numbers, $\omega = \{0, 1, 2, \dots\}$ is a transitive set. To see this, simply observe that if $n \in \omega$ and $m \in n$, then $m \in \omega$.

Exercise 8.8: Assuming that a is a transitive set, determine if each of the following sets must also be transitive. If so, explain why. If not, provide a counterexample.

1. $\{a\}$

_____

_____

_____

2. $\mathcal{P}(a)$

_____

_____

_____

3. (a, a)

_____

_____

_____

4. $\bigcup a$

_____

_____

_____

Ordinals

We are now ready to give the formal definition of an ordinal.

A set x is an **ordinal** if and only if x is well-ordered by $\in$ and x is transitive.

We generally use letters from the beginning of the Greek alphabet such as α (alpha), β (beta), and γ (gamma) to represent arbitrary ordinals. We use ω (omega) to represent the set of natural numbers when we are thinking of this set as an ordinal.

Example 8.9:

1. By part 2 of Example 8.3, every natural number is a well-ordered by $\in$. By part 6 of Example 8.7, every natural number is transitive. Therefore, every natural number is an ordinal. The natural numbers are the **finite ordinals** with 0 being the least ordinal, 1 being the next ordinal after 0, and so on.

$$0 \in 1 \in 2 \in 3 \in \cdots$$

Except for the ordinal 0, each natural number is a **successor ordinal**. A successor ordinal has the form $\alpha \cup \{\alpha\}$ for some ordinal α.

For example, $0 \cup \{0\} = \{0\} = 1$. So, 1 is a successor ordinal. It is the successor of 0.

Similarly, 2 is a successor ordinal because $1 \cup \{1\} = \{0\} \cup \{1\} = \{0, 1\} = 2$.

In general, for $n > 0$, we have $n \cup \{n\} = \{0, 1, \ldots, n - 1\} \cup \{n\} = \{0, 1, \ldots, n\} = n + 1$, showing that $n + 1$ is a successor ordinal.

Note that we previously defined the successor of the natural number n to be $n^+ = n \cup \{n\}$. The reasoning above shows that $n + 1 = n^+$. In words, the natural number $n + 1$ is the successor of the natural number n.

2. By part 3 of Example 8.3, the set of natural numbers, $\omega = \{0, 1, 2, \dots\}$ is well-ordered by $\in$. By part 7 of Example 8.7, ω is a transitive set. Therefore, ω is an ordinal. ω is the **least infinite ordinal**. It is the set containing all the finite ordinals. Note that ω is **not** a successor ordinal. Indeed, for all $n \in \omega$, $n \cup \{n\} = n + 1 \in \omega$. We say that ω is a **limit ordinal**. In fact, ω is the **least limit ordinal**.

3. If α is an ordinal, then the **successor** of α is $\alpha + 1 = \alpha \cup \{\alpha\}$. The successor of an ordinal is always an ordinal.

    For example, we have $\omega + 1 = \omega \cup \{\omega\} = \{0, 1, 2, \dots, \omega\}$. More generally, for each $n \in \omega$, we have $\omega + n = \{0, 1, 2, \dots, \omega, \omega + 1, \omega + 2, \dots, \omega + (n - 1)\}$ is an ordinal.

4. Any nonzero ordinal that is not a successor ordinal is called a **limit ordinal**. For example, we saw in part 2 above that ω is a limit ordinal. The next limit ordinal after ω is the ordinal
$$\omega + \omega = \{0, 1, 2, \dots, \omega, \omega + 1, \omega + 2, \dots\}$$

5. Define ω_1 to be the set of countable ordinals. Then ω_1 is the least uncountable limit ordinal.

6. Define **ON** to be the **class** of all ordinals. So, **ON** $= \{\alpha \mid \alpha \text{ is an ordinal}\}$. Although **ON** satisfies all the properties of being an ordinal, it is too large to be a set. Indeed, if **ON** were an ordinal, we would have **ON** $\in$ **ON**. This contradicts the fact that $\in$ is a **strict** linear ordering. **ON** is a proper class. Notice that there is no "bounding set" in the definition of **ON**, and so, the Bounded Comprehension Schema cannot be used to show that **ON** is a set.

The next fact is arguably the most important result about ordinals. It essentially says that every well-ordered set "looks" exactly like one and only one ordinal. It allows us to take the class of ordinals to be the natural representatives for all well-ordered sets.

Ordinal Fact 1: Given any well-ordered set $(A, <)$, there is a unique ordinal α such that $A \cong \alpha$.

Exercise 8.10: For each of the following well-ordered sets, find an ordinal that is isomorphic to it.

1. $(\{a, b, c, d\}, \prec)$, where $\prec$ orders $\{a, b, c, d\}$ alphabetically   _____

2. $(\mathbb{N}, <')$ (from part 4 of Example 8.1)   _____

3. $(\mathbb{N}, <_*)$ (from part 4 of Example 8.3)   _____

Ordinal Arithmetic

We will now consider the operations of addition, multiplication, and exponentiation on the ordinals.

We define the **sum** $\alpha + \beta$ of the ordinals α and β as follows:

(i)   If $\beta = 0$, then $\alpha + \beta = \alpha$.

(ii)   If $\beta = \gamma + 1$, then $\alpha + \beta = (\alpha + \gamma) + 1$.

(iii)   If β is a limit ordinal, then $\alpha + \beta = \bigcup\{\alpha + \gamma \mid \gamma \in \beta\}$.

Note: This is an example of a **definition by transfinite recursion**. Transfinite recursion will *not* be discussed in this book. For a detailed presentation of transfinite recursion, see for example, *Set Theory for Beginners*.

Example 8.11:

1. To add two natural numbers, only the first two clauses in the definition of addition are needed.

   For example, by clause (i), we have $2 + 0 = 0$.

   By clause (ii), $0 + 2 = 0 + (1 + 1) = (0 + 1) + 1 = (0 \cup \{0\}) + 1 = \{0\} + 1 = 1 + 1 = 1 \cup \{1\} = \{0\} \cup \{1\} = \{0, 1\} = 2$.

   As one more example, by clause (ii), $1 + 2 = 1 + (1 + 1) = (1 + 1) + 1 = 2 + 1$(by the same reasoning used in the above computation)$= 2 \cup \{2\} = \{0, 1\} \cup \{2\} = \{0, 1, 2\} = 3$.

2. $1 + \omega = \bigcup\{1 + n \mid n \in \omega\} = \omega$ because

   $$\bigcup\{1 + n \mid n \in \omega\} = 1 \cup 2 \cup 3 \cup \cdots = \{0\} \cup \{0, 1\} \cup \{0, 1, 2\} \cup \cdots = \{0, 1, 2, \ldots\} = \omega.$$

   We can think of $\alpha + \beta$ as follows: If we have α almonds and β berries, we can visualize $\alpha + \beta$ by placing the α almonds followed by the β berries.

   For $1 + \omega$, we place one almond followed by ω berries. The result looks just like the ordering of the natural numbers (after placement, we no longer make a distinction between almonds and berries).

$$\diamond \circ \circ \circ \circ \circ \circ \circ \circ \circ \circ \circ \circ \circ \cdots$$

Exercise 8.12: Is $\omega + 1 = 1 + \omega$? Why or why not?

_____

_____

_____

We define the **product** $\alpha\beta$ of the ordinals α and β as follows:

   (i)   If $\beta = 0$, then $\alpha\beta = 0$.
   (ii) If $\beta = \gamma + 1$, then $\alpha\beta = \alpha\gamma + \alpha$.
   (iii) If β is a limit ordinal, then $\alpha\beta = \bigcup\{\alpha\gamma \mid \gamma \in \beta\}$.

Note: This is another example of a **definition by transfinite recursion**.

Example 8.13:

1. As with addition, to multiply two natural numbers, we need only the first two clauses in the definition of multiplication.

   For example, by clause (i), we have $1 \cdot 0 = 0$.

   By clause (ii), $0 \cdot 1 = 0(0 + 1) = 0 \cdot 0 + 0 = 0 + 0$ (by clause (i))$= 0$ (by the definition of addition).

   As another example, by clause (ii), $3 \cdot 1 = 3(0 + 1) = 3 \cdot 0 + 3 = 0 + 3 = 3$.

   As one more example, $3 \cdot 2 = 3(1 + 1) = 3 \cdot 1 + 3 = 3 + 3 = 6$.

2. $2\omega = \bigcup\{2n \mid n \in \omega\} = \omega$ because

$$\bigcup\{2n \mid n \in \omega\} = 0 \cup 2 \cup 4 \cup \cdots = \emptyset \cup \{0,1\} \cup \{0,1,2,3\} \cup \cdots = \{0,1,2,\dots\} = \omega.$$

We can think of $\alpha\beta$ as follows: If we have α almonds in each of β many bags, we can visualize $\alpha\beta$ by lining up the β bags, while leaving the α many almonds in order in each bag.

For 2ω, we have ω many bags, each with 2 almonds. When we line them up, the result looks just like the ordering of the natural numbers (after placement, we no longer pay attention to the bags).

3. $\omega \cdot 2 = \omega \cdot 1 + \omega = \omega + \omega = \bigcup\{\omega + n \mid n \in \omega\} = \{0, 1, 2, \dots, \omega, \omega + 1, \omega + 2, \dots\}$. This is the next limit ordinal after ω. Notice that $\omega \cdot 2 \neq 2\omega$. Indeed, $\omega \cdot 2$ is the second limit ordinal, whereas $2\omega = \omega$ is the least limit ordinal.

This time we have just 2 bags, each with ω many almonds. This ordering looks like the natural numbers followed by a second copy of the natural numbers.

Similarly, we have $\omega \cdot 3 = \{0, 1, 2, \dots, \omega, \omega + 1, \omega + 2, \dots, \omega + \omega, \omega + \omega + 1, \omega + \omega + 2, \dots\}$. This is the next limit ordinal after $\omega \cdot 2$. Notice that $\omega \cdot 3 \neq 3\omega$. Indeed, $\omega \cdot 3$ is the third limit ordinal, whereas $3\omega = \omega$ is the least limit ordinal. $\omega \cdot 3$ can be visualized as 3 bags in a row, each bag filled up with ω many almonds all lined up in order.

In the figure above, I used three solid rectangles to indicate three bags, each containing ω many almonds, all lined up in a row.

Continuing in this fashion, we see that $\omega \cdot n$ is the nth limit ordinal. It can be visualized as n bags in a row, each bag filled up with ω many almonds, all lined up in order.

4. $\omega \cdot \omega = \bigcup\{\omega \cdot n \mid n \in \omega\}$. This is the limit of the limit ordinals of the form $\omega \cdot n$ for $n \in \mathbb{N}$. We can visualize $\omega \cdot \omega$ as ω bags, each bag filled up with ω many almonds.

Exercise 8.14: Place the following ordinals in order from least to greatest.

$$5\omega + 3 \qquad \omega \cdot 5 + 3 \qquad 3 + 5\omega \qquad 3 + \omega \cdot 5$$

$$\underline{\hspace{3cm}} \in \underline{\hspace{3cm}} \in \underline{\hspace{3cm}} \in \underline{\hspace{3cm}}$$

We define the **power** α^β of the ordinals α and β as follows:

(i) If $\beta = 0$, then $\alpha^\beta = 1$.

(ii) If $\beta = \gamma + 1$, then $\alpha^\beta = \alpha^\gamma \cdot \alpha$.

(iii) If $\beta = \bigcup\beta$, then $\alpha^\beta = \bigcup\{\alpha^\gamma \mid \gamma \in \beta\}$.

Note: This is another example of a **definition by transfinite recursion**.

Example 8.15:

1. As with addition and multiplication, we need only the first two clauses in the definition of power.

   For example, by clause (i), we have $1^0 = 1$.

   By clause (ii), $1^1 = 1^0 \cdot 1 = 1 \cdot 1 = 1$.

   As another example, by clause (ii), $2^1 = 2^0 \cdot 2 = 1 \cdot 2 = 2$.

   As one more example, $2^2 = 2^1 \cdot 2 = 2 \cdot 2 = 4$.

2. We have $\omega^0 = 1, \omega^1 = \omega^0 \cdot \omega = 1 \cdot \omega = \omega$, and $\omega^2 = \omega^1 \cdot \omega = \omega \cdot \omega$. We already saw a visual representation of $\omega^2 = \omega \cdot \omega$ in part 4 of Example 8.13.

   The reader may want to explore possible visualizations of ω^n for $n > 2$ and ω^ω. Be careful here. You may find it useful to look at ordinals such as $\omega^2 \cdot 2, \omega^2 \cdot 3,\dots$ and so on, to help understand ω^3. And similarly, for higher powers of ω.

3. $2^\omega = \bigcup\{2^n \mid n \in \omega\} = 2 \cup 4 \cup 8 \cup 16 = \omega$. More generally, $k^\omega = \omega$ for every $k \in \omega$.

Note: Readers that are somewhat familiar with cardinal arithmetic might be very suspicious of the equation $2^\omega = \omega$. Isn't 2^ω uncountable? Unfortunately, the standard definitions for ordinal and cardinal arithmetic are very different from each other. We will be looking at cardinals and cardinal exponentiation next.

Cardinals

In order to discuss cardinals, it is important that we assume AC (the axiom of choice). AC is equivalent to the following axiom:

The **Well Ordering Axiom (WA):** Every set can be well-ordered.

Assuming AC (or equivalently WA), we can define the **cardinality** of a set A, written $|A|$, to be the least ordinal α such that $(A, <) \cong \alpha$, where $<$ is some well ordering of A.

An ordinal α is a **cardinal** if and only if $|\alpha| = \alpha$.

We generally use letters from the middle of the Greek alphabet such as κ (kappa), λ (lambda), and μ (mu) to represent cardinals.

Note: If α and β are ordinals, we may write $\alpha < \beta$ or $\beta > \alpha$ instead of $\alpha \in \beta$. $\alpha \leq \beta$ is an abbreviation for $\alpha \in \beta$ or $\alpha = \beta$. Similarly, $\beta \geq \alpha$ has the same meaning as $\alpha \leq \beta$. We are most likely to use $<$ and $\leq$ (instead of $\in$) when comparing cardinals. However, when comparing cardinals (or ordinals), $<$ and $\in$ have the same exact meaning.

Example 8.16:

1. Each natural number $n \in \omega$ is a cardinal, as there is no bijection from a natural number to a smaller natural number.

2. ω is a cardinal because ω is infinite, while any smaller ordinal is finite.

   Many authors use the Hebrew letter $\aleph$ (pronounced "aleph") with a subscript of 0 to represent the least infinite cardinal. In other words, when thinking of ω as a cardinal, $\aleph_0$ (pronounced "aleph naught") is often used. We will continue to use ω. For all practical purposes, $\mathbb{N}$, ω, and $\aleph_0$ all represent the same set with the same natural ordering.

3. $\omega_1 = \{\alpha \mid \alpha$ is a countable ordinal$\}$ is a cardinal because ω_1 is uncountable, while any smaller ordinal is countable. Many authors use $\aleph_1$ (pronounced "aleph one") to represent the least uncountable cardinal. We will continue to use ω_1 for this purpose. As in 2 with ω, for all practical purposes, ω_1 and $\aleph_1$ both represent the same set with the same ordering.

Exercise 8.17: Determine if each of the following ordinals is a cardinal. If so, explain why. If not, describe a bijection between the given ordinal and a smaller ordinal.

1. 5

2. $\omega + 1$

3. $1 + \omega$

4. $\omega \cdot 2$

5. 2ω

If α is an ordinal, then α^+ is the least cardinal such that $\alpha \in \alpha^+$.

Example 8.18:

1. If n is a natural number, then the definition of n^+ given above agrees with the definition from the end of Lesson 2. In other words, for $n \in \omega$, we have $n^+ = n + 1$.

2. $\omega^+ = \omega_1$.

3. $(\omega + 1)^+ = \omega_1$.

4. For any ordinal α with $\omega \in \alpha \in \omega_1$, $\alpha^+ = \omega_1$.

A cardinal κ is a **successor cardinal** if and only if there is an ordinal α such that $\kappa = \alpha^+$. Any infinite cardinal that isn't a successor cardinal is called a **limit cardinal**.

Example 8.19:

1. Every nonzero natural number is a successor cardinal.

2. ω is a limit cardinal.

3. 0 is the only cardinal that is neither a successor cardinal nor a limit cardinal.

4. We define $\omega_2 = \omega_1^+$, $\omega_3 = \omega_2^+$,..., and in general, $\omega_{n+1} = \omega_n^+$. By definition, the cardinals $\omega_1, \omega_2, \omega_3, \dots$ are all successor cardinals.

Exercise 8.20: Let $\omega_\omega = \bigcup\{\omega_n | n \in \omega\}$.

1. Explain why ω_ω is a cardinal.

_____

_____

_____

2. Explain why ω_ω is a limit cardinal.

_____

_____

_____

Cardinal Arithmetic

We define addition, multiplication, and exponentiation of cardinals as follows:

$$\kappa + \lambda = |(\kappa \times \{0\}) \cup (\lambda \times \{1\})| \qquad \kappa \cdot \lambda = |\kappa \times \lambda| \qquad \kappa^\lambda = |{}^\lambda\kappa|$$

Notes: (1) The choice of using $\kappa \times \{0\}$ and $\lambda \times \{1\}$ in the definition of $\kappa + \lambda$ is somewhat arbitrary. We can define $\kappa + \lambda$ to be $|A \cup B|$, where A and B are **any** two **disjoint** sets such that $|A| = \kappa$ and $|B| = \lambda$.

(2) Recall that $\kappa \times \lambda$ is the **Cartesian product** of κ and λ. See Lesson 3 for details.

(3) Recall that $^\lambda\kappa$ is the set of functions from λ to κ. See Lesson 5 for details.

Example 8.21:

1. $\omega + \omega = |A \cup B|$, where $A = \omega \times \{0\}$ and $B = \omega \times \{1\}$. Since A and B are both countably infinite, $A \cup B$ is countably infinite (see Problem 68 in Problem Set 6). Therefore, $|A \cup B| = \omega$. So, $\omega + \omega = \omega$.

2. $\omega \cdot \omega = |\omega \times \omega| = \omega$ by Problem 22 in Problem Set 6.

3. For any cardinal κ, $\kappa^0 = |\{f \mid f : 0 \to \kappa\}| = |0| = 0$.

4. For any cardinal κ, $\kappa^1 = |\{f \mid f : \{0\} \to \kappa\}| = |\kappa| = \kappa$.

5. $2^\omega = |^\omega 2| = |\{f \mid f : \omega \to \{0,1\}\}|$. Some other sets that have the same size as 2^ω are $\mathcal{P}(\omega)$, $\mathbb{R}$, and ω^ω (for example, see Problem 52 in Problem Set 6). We also know by Cantor's Theorem (Equinumerosity Fact 1 from Lesson 6) that 2^ω is strictly larger than ω. In particular, 2^ω is uncountable.

   Exactly how big is 2^ω? The **Continuum Hypothesis (CH)** is the statement $2^\omega = \omega_1$. In other words, CH says that 2^ω is the least uncountable cardinal. This is equivalent to saying that there is no set A such that $|\mathbb{N}| < |A| < |\mathbb{R}|$. As it turns out, CH is **independent** of the axioms of ZFC. It can neither be proved nor disproved using only the axioms we described in Lesson 7. Showing that CH is independent of ZFC is beyond the scope of this book. It is best left for the interested reader to investigate "independently."

In order to be able to do basic cardinal arithmetic, the following fact is extremely important.

Cardinal Fact 1: If κ is an infinite cardinal, then $\kappa \cdot \kappa = \kappa$.

Exercise 8.22: Let κ be an infinite cardinal. Explain why $\kappa + \kappa = \kappa$ (**Hint:** use Cardinal Fact 1).

_____

_____

_____

Cofinality

Let α and β be ordinals. A function $f : \alpha \to \beta$ is **cofinal** if and only if for every $\delta \in \beta$, there is $\gamma \in \alpha$ such that $\delta \leq f(\gamma)$.

In other words, $f : \alpha \to \beta$ cofinal means that the range of f is unbounded in β. In simple terms, a cofinal function "cuts" all the way through its target ordinal.

The **cofinality** of an ordinal β, written $cf(\beta)$, is the least ordinal α such that there exists a cofinal function $f : \alpha \to \beta$.

Example 8.23:

1. If β is a successor ordinal, then $cf(\beta) = 1 = \{0\}$. To see this, note that there is an ordinal γ such that $\beta = \gamma + 1$. Define $f: \{0\} \to \beta$ by $f(0) = \gamma$. Then f is cofinal.

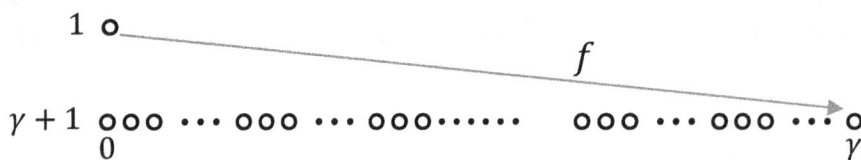

2. For any ordinal β, the identity function $i: \beta \to \beta$ defined by $i(\gamma) = \gamma$ for all $\gamma \in \beta$ is cofinal. It follows that $cf(\beta) \leq \beta$.

3. If n is a natural number, then there is no cofinal function from n to ω. So, by 2, $cf(\omega) = \omega$.

   ω is an example of a **regular** cardinal. In general, a cardinal κ is called regular if $cf(\kappa) = \kappa$.

4. ω_1 is also a regular cardinal. To see this, let $f: \alpha \to \omega_1$ be a cofinal map. Then we have $\omega_1 = \cup\{f(\beta)|\beta \in \alpha\}$. If $\alpha < \omega_1$, then this is a countable union of countable sets, and so, by Problem 68 in Problem Set 6, ω_1 would be countable. Since ω_1 is uncountable, $\omega_1 \leq \alpha$. It follows that $cf(\omega_1) = \omega_1$.

   In fact, every infinite successor cardinal is regular. You will be asked to explain why this is true in Problem 81 below.

   In particular, the following cardinals are all regular: $\omega_1, \omega_2, \omega_3,...$ and so on.

5. An infinite cardinal that is **not** regular is called **singular**. Since $cf(\kappa) \leq \kappa$, an infinite cardinal κ is singular if and only if $cf(\kappa) < \kappa$.

   For example, ω_ω is singular. We can define a cofinal map $f: \omega \to \omega_\omega$ by $f(n) = \omega_n$. It follows that $cf(\omega_\omega) \leq \omega$. In fact, it's easy to see that $cf(\omega_\omega) = \omega$, as there can certainly be no cofinal map from a natural number to ω_ω, simply because ω_ω is infinite.

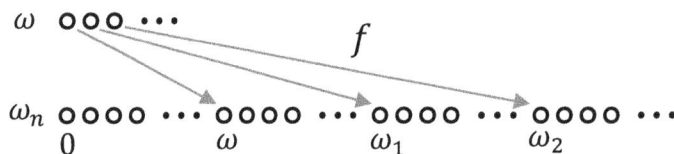

In part 5 of Example 8.21, we discussed what we know about the size of 2^ω. The axioms of ZFC imply that $2^\omega > \omega$. If CH holds, then we know that the exact value of 2^ω is ω_1, the first uncountable cardinal. As we already mentioned, CH is independent of ZFC. So, it is consistent with ZFC that $2^\omega = \omega_1$ and it is also consistent with ZFC that $2^\omega \neq \omega_1$. So, the next natural question is, "For which uncountable cardinals κ is it consistent with ZFC that $2^\omega = \kappa$?" As it turns out, the possible values for κ for which $2^\omega = \kappa$ is unbounded in **ON**. So, not only can we not determine the size of 2^ω using the axioms of ZFC, but we cannot even put a bound on the size of 2^ω.

We will not attempt here to determine all the values κ for which it is consistent that $2^\omega = \kappa$. However, using the next fact, we will be able to determine certain values of κ such that $2^\omega \neq \kappa$.

Specifically, we will see that $cf(2^\omega) > \omega$. So, for example, since $cf(\omega_\omega) = \omega$, $2^\omega \neq \omega_\omega$.

Cofinality Fact 1: If κ is an infinite cardinal, then $\kappa < \kappa^{cf(\kappa)}$ (this result is known as **König's Theorem**).

Exercise 8.24: Explain why $2^{\omega} \neq \omega_{\omega}$ (**Hint:** use Cofinality Fact 1).

_____

_____

_____

Problem Set 8

LEVEL 1

Determine if each of the following ordered sets is well-ordered. Note that $<$ is always the usual order on the given set.

1. $(\mathbb{Z}^+, <)$

2. $((0, 1), <)$

3. $([1, 8), <)$

4. $([1, 8) \cap \mathbb{Q}, <)$

5. $(\{0, 1, 2, 3, 4\}, <)$

6. $(2\mathbb{N}, <)$

7. $(X, <_D)$, where $X = \{x \mid x$ is a word in the English language$\}$ and $<_D$ is the dictionary order on X.

Find an ordinal isomorphic to the given well-ordered set.

8. $(2\mathbb{N} + 1, <)$

9. $(\{a, b, c, d, e, f, g\}, <)$, where $<$ orders the given set alphabetically

10. $(A, <^\star)$ (from Exercise 8.2)

11. $(\mathbb{Z}^-, >)$

12. $\text{pred}(\mathbb{N}, 0)$

13. $\text{pred}(\mathbb{N}, 30)$

14. $\text{pred}(A, \star)$, where $(A, <^\star)$ is as defined in Exercise 8.2

Use cardinal arithmetic to compute each of the following:

15. $5 + 11$

16. $7 + \omega$

17. $\omega + 7$

18. $\omega + \omega$

19. $\omega \cdot \omega$

20. $10 \cdot \omega$

LEVEL 2

Use ordinal arithmetic to write each of the following ordinals in the form $\omega \cdot a + b$, where $a, b \in \omega$.

21. $7 + 3$

22. $7 + \omega$

23. $\omega + 7$

24. 7ω

25. $\omega \cdot 7$

26. $7\omega + 7$

27. $\omega \cdot 7 + 7$

28. $7 + 7\omega$

29. $7 + \omega \cdot 7$

30. $7\omega + \omega$

31. $7\omega + 7\omega$

32. $\omega \cdot 7 + \omega$

33. $\omega + \omega \cdot 7$

34. $\omega \cdot 7 + \omega \cdot 7$

Determine if each of the following statements is true or false. If true, explain why. If false, provide a counterexample.

35. Every well-ordered set is isomorphic to an initial segment of the set of natural numbers.

36. If $(A, <)$ is an infinite well-ordered set, then ω is isomorphic to an initial segment of $(A, <)$.

37. Every transitive set is an ordinal.

38. Every set that is well-ordered by $\in$ is an ordinal.

39. If α, β, and γ are ordinals with $\alpha \in \beta$ and $\beta \in \gamma$, then $\alpha \in \gamma$.

40. There exist two well-ordered sets such that neither one is isomorphic to an initial segment of the other one.

LEVEL 3

Determine if each of the following ordinals is a cardinal. If so, explain why. If not, describe a bijection between this ordinal and a smaller ordinal. Use ordinal arithmetic where appropriate.

41. 0

42. 15

43. $\omega + 7$

44. $7 + \omega$

45. $\omega \cdot 7$

46. 7ω

47. ω^2

48. $5\omega^2 + 4$

Use ordinal arithmetic to write each of the following ordinals in the form $\omega^2 \cdot a + \omega \cdot b + c$, where $a, b \in \omega$.

49. $2\omega \cdot \omega$

50. $(\omega \cdot \omega) \cdot 2$

51. $\omega^2 \cdot 3 + 5\omega$

52. $5\omega + 3\omega^2$

53. $5\omega + \omega^2 \cdot 3$

54. $\omega \cdot 5 + \omega^2 \cdot 3$

Let κ and λ be infinite cardinals. Verify each of the following:

55. $\kappa + \lambda = \max\{\kappa, \lambda\}$.

56. $\kappa \cdot \lambda = \max\{\kappa, \lambda\}$.

57. $\kappa^3 = \kappa$.

58. For each natural number $n \geq 1$, $\kappa^n = \kappa$.

LEVEL 4

Let κ, λ, and μ be infinite cardinals. Verify each of the following:

59. If $\mu \leq \lambda$, then $\kappa^\mu \leq \kappa^\lambda$.

60. $\kappa^\omega \leq \kappa^\lambda$.

61. $\kappa^\omega \leq \kappa^\mu$.

62. If $\mu \leq \lambda$, then $\kappa^\lambda . \kappa^\mu = \kappa^\lambda$.

63. If $\mu \leq \lambda$, then $\kappa^{\lambda+\mu} = \kappa^\lambda$.

64. $\kappa^\lambda \cdot \kappa^\mu = \kappa^{\lambda+\mu}$.

Determine if each of the following statements is true or false. If true, explain why. If false, provide a counterexample.

65. If α is an ordinal, then $\in$ is a transitive relation on $\alpha + 1 = \alpha \cup \{\alpha\}$.

66. If α is an ordinal, then $\in$ is a trichotomous relation on $\alpha + 1 = \alpha \cup \{\alpha\}$.

67. If α is an ordinal, then $\alpha + 1 = \alpha \cup \{\alpha\}$ is well-ordered by $\in$.

68. If α is a transitive set, then $\alpha + 1 = \alpha \cup \{\alpha\}$ is also a transitive set.

69. If α is an ordinal, then $\alpha + 1 = \alpha \cup \{\alpha\}$ is also an ordinal.

70. If α is an ordinal and $x \in \alpha$, then x is an ordinal.

71. If κ a cardinal and $x \in \kappa$, then x is a cardinal.

72. If α is an ordinal and $x \in \alpha$, then $x = \mathrm{pred}(\alpha, x)$.

LEVEL 5

Let X be a nonempty set of ordinals. Verify each of the following:

73. $\bigcup X$ is a transitive set.

74. $\bigcup X$ is well-ordered by $\in$.

75. If $\alpha \in X$, then $\alpha \in \bigcup X$.

76. $\bigcup X$ is the least ordinal greater than or equal to all elements of X.

77. $\bigcap X$ is an ordinal.

78. $\bigcap X$ is the least ordinal in X.

79. If every ordinal in X is a cardinal, then $\bigcup X$ is a cardinal.

Determine if each of the following statements is true or false. If true, explain why. If false, provide a counterexample.

80. Every infinite cardinal is a limit ordinal.

81. Every infinite successor cardinal is regular.

82. Every limit cardinal is regular.

83. Show that a transitive set of ordinals in an ordinal.

84. Let κ be an infinite cardinal and for each ordinal $\alpha < \kappa$ let X_α be a set such that $|X_\alpha| \leq \kappa$. Show that $|\bigcup\{X_\alpha \mid \alpha < \kappa\}| \leq \kappa$.

85. Show that for any ordinal α, $cf\big(cf(\alpha)\big) = cf(\alpha)$.

86. The **Generalized Continuum Hypothesis (GCH)** is the statement $\forall \kappa \geq \omega(2^\kappa = \kappa^+)$. Assuming GCH, compute κ^λ for all cardinals λ and κ.

Lesson 1

Exercise 1.2:

1. 5 elements; j, k, t, u and v.

2. 3 elements; triceratops, tulip, and lima bean

3. 4 elements; $1.6, 5.66, 9.03,$ and 15.27

4. 6 elements; Earth, Mars, Neptune, Venus, Jupiter, and Mercury

5. 4 elements; oxygen, helium, nitrogen, and sulfer

Exercise 1.4:

1. $\{5, 7, 9\}$, $\{7, 5, 9\}$, $\{3, 3, 7, 9\}$ $\{5, 5, 7, 9, 9\}$

2. $\{a, x, t, v, d, b\}$ $\{a, a, x, x, x, t, t, t, t, v, v, v, d, d\}$, $\{t, a, x, d, v, d, x, a, t\}$, $\{t, a, x, x, v, d\}$

3. $\{\text{car}, \text{bus}\}$, $\{c, a, r, b, u, s\}$ $\{\text{bus}, \text{car}, \text{bus}\}$, $\{\text{car}, \text{car}, \text{bus}, \text{car}\}$

Exercise 1.6:

1. This is **true** because 8 is an element of Y.

2. This is **false** because m is an element of Y.

3. This is **false** because leopard is **not** an element of Y.

4. This is **false** because 6 is **not** an element (or member) of Y.

5. This is **true** because y is **not** an element of Y.

6. This is **true** because 101 and v are both elements of Y.

7. This is **true** because $1, 8, 101,$ and jaguar are all elements of Y.

8. This is **false** because a is **not** an element of Y.

Exercise 1.8: $\{5, 7, 9, 11, 13, 15, 17, 19, 21, 23, 25, 27, 29, 31, 33\}$

Exercise 1.10:

1. $2\mathbb{Z} = \{\ldots, -6, -4, -2, 0, 2, 4, 6, \ldots\}$

2. $2\mathbb{Z} + 1 = \{\ldots, -5, -3, -1, 1, 3, 5, 7, \ldots\}$

3. $\mathbb{Z}^- = \{\ldots, -6, -5, -4, -3, -2, -1\}$

Exercise 1.12:

1. **No** because the word "hearth" contains just two vowels: e and a.

2. **Yes** because the word "grasshopper" contains the three distinct vowels a, o, and e.

3. **No** because the phrase "fire away" consists of two words and not a single word.

4. **Yes** because the word "incremental" contains the three distinct vowels i, e, and a.

5. **No** because the word "apartment" contains just two distinct vowels: a and e. Note that in this example, the word contains three vowels (because the a appears twice), but only two **distinct** vowels.

Exercise 1.14: Note that there are many different correct descriptions for each of the sets given. I provide just one of each here.

1. $\{n \mid n \text{ is an even natural number and } 0 \leq n \leq 16\}$

2. $\{n \mid n \text{ is an odd integer between } -5 \text{ and } 87, \text{ inclusive}\}$

3. $\{n \mid n \text{ is a natural number}\}$

4. $\{n \mid n \text{ is an integer}\}$

5. $\{n \mid n \text{ is an even integer}\}$

Exercise 1.16: $\frac{3}{5}, \frac{6}{10}$ $\frac{-13}{7}, \frac{13}{-7}$ $\frac{0}{6}, 0, \frac{0}{-1}$ $\frac{4}{4}, \frac{-17}{-17}, 1, \frac{1}{1}$ $\frac{-5}{-3}, \frac{20}{12}$ $\frac{10}{-6}, \frac{-15}{9}$

Exercise 1.17:

1. **rational**

2. **rational**

3. **irrational**

4. **rational**

5. **irrational**

Exercise 1.18: See the image to the right.

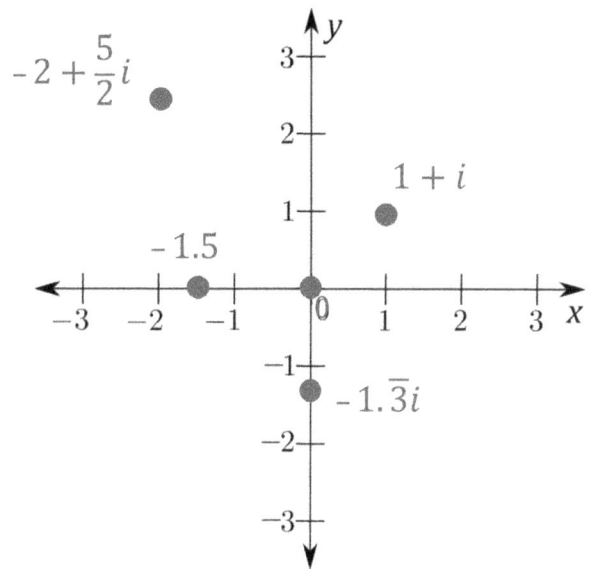

Exercise 1.20:

1. 3 elements; $\{a\}, \{b, c\}$, and $\{d, e, f\}$.

2. 4 elements; $\{x\}, \{x, \{x\}\}, \{\{x\}\}$, and $\left\{x, \{x\}, \{\{x\}\}\right\}$

3. 1 element $\left\{\emptyset, \{\emptyset\}, \{\emptyset, \{\emptyset\}\}\right\}$

4. 1 element; $\{z\}$

Note: For number 4 above, we have $\{z, z\} = \{z\}$ and $\{z, z, z\} = \{z\}$. Therefore, it follows that $\{\{z\}, \{z, z\}, \{z, z, z\}\} = \{\{z\}, \{z\}, \{z\}\} = \{\{z\}\}$. So, $\{z\}$ is the only element of $\{\{z\}, \{z, z\}, \{z, z, z\}\}$.

Exercise 1.22:

1. $|\{1, 2, 3, \ldots, 80\}| = \mathbf{80}$

2. $|\{c, d, e, f, e, d, c\}| = \mathbf{4}$ (note that $\{c, d, e, f, e, d, c\} = \{c, d, e, f\}$)

3. $\left|\{\emptyset, \{\emptyset\}, \{\emptyset, \{\emptyset\}\}\}\right| = \mathbf{3}$ (the three elements of the set are $\emptyset$, $\{\emptyset\}$, and $\{\emptyset, \{\emptyset\}\}$)

4. $|\{n \in \mathbb{N} \mid 222 \leq n \leq 2222\}| = 2222 - 222 + 1 = \mathbf{2001}$ (use the fence-post formula)

5. $\{x, x\} = \{x\}$, $\{x, x, x\} = \{x\}$, and $\{x, \{x, x\}\} = \{x, \{x\}\}$.

   Therefore, $\left\{x, \{x, x\}, \{x, x, x\}, \{x, \{x, x\}\}\right\} = \left\{x, \{x\}, \{x, \{x\}\}\right\}$.

   So, $\left|\left\{x, \{x, x\}, \{x, x, x\}, \{x, \{x, x\}\}\right\}\right| = \left|\left\{x, \{x\}, \{x, \{x\}\}\right\}\right| = \mathbf{3}$. (the three elements of the set are $x, \{x\}$, and $\{x, \{x\}\}$)

Exercise 1.24:

1. $\boldsymbol{B \subseteq A}$. Also, $A \nsubseteq B$ because $b \in A$, whereas $b \notin B$.

2. **Neither.** $A \nsubseteq B$ because $-2 \in A$, whereas $-2 \notin B$. $B \nsubseteq A$ because $1 \in B$, whereas $1 \notin A$.

3. **Both.** $A = \{x, \{x, x\}\} = \{x, \{x\}\}$ and $B = \{\{x\}, x, x\} = \{\{x\}, x\} = \{x, \{x\}\}$. So, $A = B$.

4. $\boldsymbol{B \subseteq A}$. Also, $A \nsubseteq B$ because $-3 \in A$, whereas $-3 \notin B$.

Exercise 1.26:

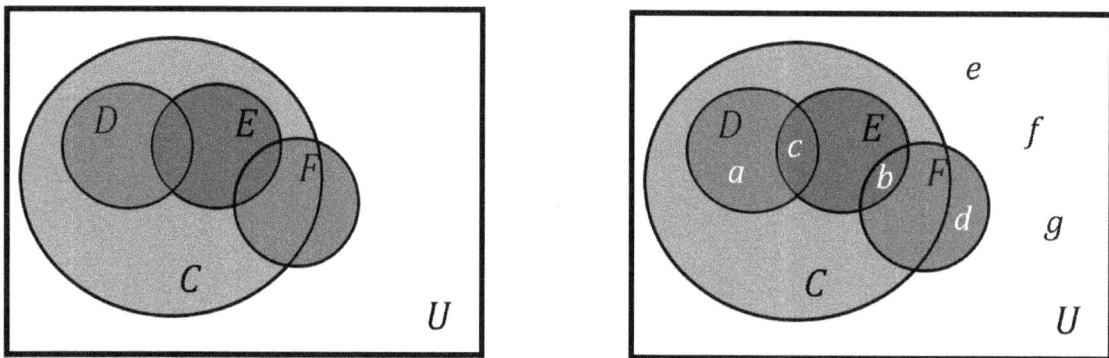

Note: Above are possible Venn diagrams for this problem. The diagram on the left shows the relationship between the sets C, D, E, and F. Notice how D and E are both subsets of C, whereas F is not a subset of C. Also, notice how D and E overlap, E and F overlap, but there is no overlap between D and F (they have no elements in common). The diagram on the right shows the proper placement of the elements. Here, I chose the universal set to be $U = \{a, b, c, d, e, f, g\}$. This choice for the universal set is somewhat arbitrary. Any set containing $\{a, b, c, d\}$ would do.

Exercise 1.28: $\{a, b, c, d\}$ has **16** subsets. We can also say that the cardinality of the power set of $\{a, b, c, d\}$ is 16. That is, $|\mathcal{P}(\{a, b, c, d\})| = \mathbf{16}$. Below is a tree diagram.

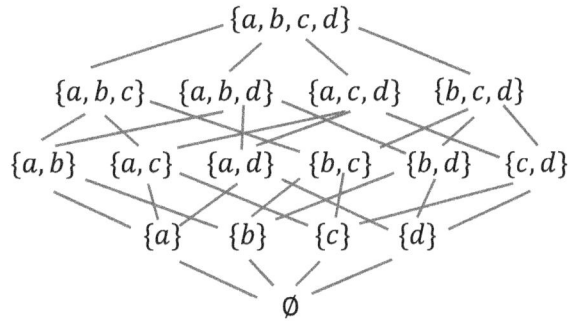

Exercise 1.30: $|\mathcal{P}(X)| = 128 = 2^7$. Therefore, $|X| = 7$.

Exercise 1.31: $4 = \{0, 1, 2, 3\} = \Big\{\emptyset, \{\emptyset\}, \{\emptyset, \{\emptyset\}\}, \big\{\emptyset, \{\emptyset\}, \{\emptyset, \{\emptyset\}\}\big\}\Big\}$

Lesson 2

Exercise 2.2:

1. $A \cup B = \{a, b, c, \Delta, \delta, \gamma\}$.

2. $A \cap B = \{b, \delta\}$.

3. $A \setminus B = \{a, \Delta\}$

4. $B \setminus A = \{c, \gamma\}$

5. $A \, \Delta \, B = (A \setminus B) \cup (B \setminus A) = \{a, \Delta\} \cup \{c, \gamma\} = \{a, c, \Delta, \gamma\}$

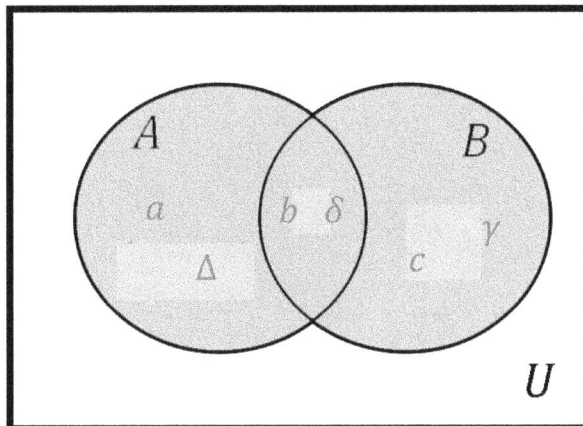

Exercise 2.5:

1. **No**. For example, $1 + 2i \in A$ because $1, 2 \in \mathbb{Z}$, but $1 + 2i \notin B$ because $1 \in \mathbb{Q}$.

2. **No**. For example, $0.1010010001 \ldots + 3i \in B$ because $0.1010010001 \ldots \notin \mathbb{Q}$, but $0.1010010001 \ldots + 3i \notin A$ because $0.1010010001 \ldots \notin \mathbb{Z}$.

3. **Yes**. To see this, suppose that $a + bi \in A \cap B$. Then $a + bi \in A$, and so, $a \in \mathbb{Z}$. Since $\mathbb{Z} \subseteq \mathbb{Q}$, $a \in \mathbb{Q}$. Also, $a + bi \in B$. So, $a \notin \mathbb{Q}$. Since we cannot have both $a \in \mathbb{Q}$ and $a \notin \mathbb{Q}$, we must have $A \cap B = \emptyset$.

Exercise 2.6:

1. $A = \{\mathbf{0}\}, B = \{\mathbf{1}\}$. Then $A \cup B = \{0,1\}$. Since $1 \in A \cup B$ and $1 \notin A$, $A \cup B \nsubseteq A$.

2. $A = \{\mathbf{0,1}\}, B = \{\mathbf{1,2}\}$. Then $A \cap B = \{1\}$. Since $0 \in A$ and $0 \notin A \cap B$, $A \nsubseteq A \cap B$.

3. $A = \{\mathbf{0,1}\}, B = \{\mathbf{0}\}$. Then $B \subseteq A$, but $A \cup B = \{0,1\} \neq \{0\} = B$.

4. $A = \{\mathbf{0,1}\}, B = \{\mathbf{0}\}$. Then $B \subseteq A$, but $A \setminus B = \{1\} \neq \emptyset$.

Exercise 2.9:

1. **Interval.**

2. **Not an interval.** $0,1 \in \mathbb{Z}$, $0 < \frac{1}{2} < 1$, but $\frac{1}{2} \notin \mathbb{Z}$.

3. **Not an interval.** $\frac{1}{2}, 1 \in G, \frac{1}{2} < \frac{3}{4} < 1$, but $\frac{3}{4} \notin G$.

4. **Interval.**

5. **Interval.**

6. **Not an interval.** $0,1 \in \mathbb{Q}$, $0 < 0.01001000100001\ldots < 1$, but $0.01001000100001\ldots \notin \mathbb{Q}$.

Exercise 2.11:

Exercise 2.13:

1. $C \cup D = (-\infty, 3]$

2. $C \cap D = (-1, 2]$

3. $C \setminus D = (-\infty, -1]$

4. $D \setminus C = (2, 3]$

5. $C \, \Delta \, D = (-\infty, -1] \cup (2, 3]$

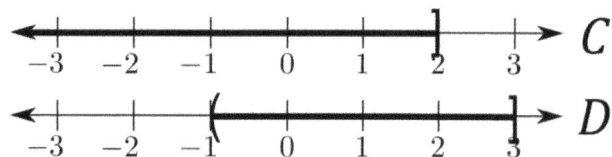

Exercise 2.15:

1. $\bigcup X = \{a,b,c,d,e,x,y,z\}$ $\bigcap X = \{b\}$

2. $\bigcup X = [-17, \infty)$ $\bigcap X = (2,3)$

3. $\bigcup X = \mathbb{Z}$ $\bigcap X = \{0\}$

Exercise 2.18:

1. $3^+ = 3 \cup \{3\} = \{0,1,2\} \cup \{3\} = \{0,1,2,3\} = \mathbf{4}$

2. $3^- = \mathbf{2}$ (because $2^+ = 3$ – see part 3 of Example 2.16)

3. $(3^+)^- = \mathbf{3}$

4. $(3^-)^+ = \mathbf{3}$

5. $4 \cup \{4\} = \{0,1,2,3\} \cup \{4\} = \{0,1,2,3,4\} = \mathbf{5}$

168

6. $4 \cap \{4\} = \{0, 1, 2, 3\} \cap \{4\} = \emptyset = 0$

7. $3 \cup 5 = \{0, 1, 2\} \cup \{0, 1, 2, 3, 4\} = \{0, 1, 2, 3, 4\} = 5$

8. $3 \cap 5 = \{0, 1, 2\} \cap \{0, 1, 2, 3, 4\} = \{0, 1, 2\} = 3$

Lesson 3

Exercise 3.2:

1. **False**

2. **False**

3. **True**

4. **False**

Exercise 3.4:

1. $x = 2, y = 7$

2. $x = 3, y = 3$

3. $x = 1, y = 5$

4. $x = 5, y = 1$ (note that $\{\{1, 5\}, \{5\}\} = \{\{5\}, \{5, 1\}\}$)

5. **No solution.**

6. **No solution.**

7. $x = 9, y = 9$ (note that $(9, 9) = \{\{9\}, \{9, 9\}\} = \{\{9\}, \{9\}\} = \{\{9\}\}$)

8. **No solution.**

Exercise 3.5:

1. $(x, y, z) = ((x, y), z) = \{\{(x, y)\}, \{(x, y), z\}\} = \left\{ \left\{ \{\{x\}, \{x, y\}\} \right\}, \left\{ \{\{x\}, \{x, y\}\}, z \right\} \right\}$

2. $(x, x, x) = ((x, x), x) = (\{\{x\}\}, x) = \left\{ \left\{ \{\{x\}\} \right\}, \left\{ \{\{x\}\}, x \right\} \right\}$

Exercise 3.7:

1. **True** (because $0 \in \mathbb{N}$)

2. **False**

3. **True** (because $(0, 0) = \{\{0\}, \{0, 0\}\} = \{\{0\}, \{0\}\} = \{\{0\}\}$)

4. **True**

5. **False**

6. **False** ($|\{a, b, c, d\} \times \{\alpha, \beta, \gamma\}| = |\{a, b, c, d\}| \cdot |\{\alpha, \beta, \gamma\}| = 4 \cdot 3 = 12$)

169

Exercise 3.9: $\{(0, 1, 0, 2), (0, 1, 1, 2), (0, 1, 2, 2), (0, 2, 0, 2), (0, 2, 1, 2), (0, 2, 2, 2)\}$

Exercise 3.11:

1. $\mathbb{R}^2 = \mathbb{R} \times \mathbb{R} = \{(x, y) \mid x, y \in \mathbb{R}\}$.

2. $\mathbb{Z}^3 = \mathbb{Z} \times \mathbb{Z} \times \mathbb{Z} = \{(x, y, z) \mid x, y, z \in \mathbb{Z}\}$.

3. $\mathbb{C}^5 = \mathbb{C} \times \mathbb{C} \times \mathbb{C} \times \mathbb{C} \times \mathbb{C} = \{(z_1, z_2, z_3, z_4, z_5) \mid z_1, z_2, z_3, z_4, z_5 \in \mathbb{C}\}$.

4. $\mathbb{R}^n = \{(x_1, x_2, \ldots, x_n) \mid x_1, x_2, \ldots, x_n \in \mathbb{R}\}$.

5. $\mathbb{C}^n = \{(z_1, z_2, \ldots, z_n) \mid z_1, z_2, \ldots, z_n \in \mathbb{C}\}$.

Exercise 3.13:

1. **False.**

2. **True.**

3. **True.**

4. **False.**

Exercise 3.15:

1. $|X \times X| = |X| \cdot |X| = 2 \cdot 2 = \mathbf{4}$

2. $|\mathcal{P}(X \times X)| = 2^4 = \mathbf{16}$

3. $\emptyset, \{(0, 0)\}, \{(0, 1)\}, \{(1, 0)\}, \{(1, 1)\},$

   $\{(0, 0), (0, 1)\}, \{(0, 0), (1, 0)\}, \{(0, 0), (1, 1)\}, \{(0, 1), (1, 0)\}, \{(0, 1), (1, 1)\}, \{(1, 0), (1, 1)\},$

   $\{(0, 0), (0, 1), (1, 0)\}, \{(0, 0), (0, 1), (1, 1)\}, \{(0, 0), (1, 0), (1, 1)\}, \{(0, 1), (1, 0), (1, 1)\},$

   $\{(0, 0), (0, 1), (1, 0), (1, 1)\}$

Exercise 3.17:

1. $\operatorname{dom} R = \{0, 1, 5, 9\}$; $\operatorname{ran} R = \{a, b, c, d\}$; $\operatorname{field} R = \{0, 1, 5, 9, a, b, c, d\}$

2. $\operatorname{dom} S = \mathbb{Z}$; $\operatorname{ran} S = \mathbb{C}$; $\operatorname{field} R = \mathbb{Z} \cup \mathbb{C} = \mathbb{C}$

3. $\operatorname{dom} T = \mathbb{Z} \times \mathbb{Z}^*$; $\operatorname{ran} T = \mathbb{Z} \times \mathbb{Z}^*$; $\operatorname{field} T = (\mathbb{Z} \times \mathbb{Z}^*) \cup (\mathbb{Z} \times \mathbb{Z}^*) = \mathbb{Z} \times \mathbb{Z}^*$

Exercise 3.19:

1. **No.** For example, $1 \not> 1$.

2. **No.** For example, $2 > 1$, but $1 \not> 2$.

3. **Yes.**

4. **Yes.**

5. **Yes.** We say that this is true **vacuously** because $a > b$ and $b > a$ cannot occur simultaneously.

6. **Yes.**

Exercise 3.21:

1. $|X^3| = |X| \cdot |X| \cdot |X| = 2 \cdot 2 \cdot 2 = \mathbf{8}$

2. $|\mathcal{P}(X^3)| = 2^8 = \mathbf{256}$

3. $|\mathcal{P}(X^4)| = 2^{16} = \mathbf{65,536}$

4. $(\mathbf{1,1,1,1})$ is the only element of R.

Exercise 3.23:

1. **No** because $\leq$ is **not** trichotomous by part 3 of Example 3.18.

2. **No** because $\leq$ is **not** trichotomous (for example, $0 \leq 0$ and $0 = 0$ both hold).

3. **Yes**.

4. **Yes**.

5. **No** because $\subset$ is **not** trichotomous: if $a, b \in A$ are distinct, then $\{a\} \not\subset \{b\}$, $\{b\} \not\subset \{a\}$, and $\{a\} \neq \{b\}$.

Exercise 3.24:

1. If $a \in A$, then $a \leq a$ and $a = a$.

2. Let $a \in A$. Since $a = a$, by definition we have $a \leq a$.

3. Let $a, b \in A$ with $a \leq b$ and $b \leq a$. If $a \neq b$, then $a < b$ and $b < a$. But this cannot happen because $<$ is trichotomous on A. So, $a = b$.

4. Let $a, b, c \in A$ with $a \leq b$ and $b \leq c$. If $a = b$, then we have $a \leq c$ by direct substitution. Similarly, if $b = c$, we have $a \leq c$ by direct substitution. If $a < b$ and $b < c$, then $a < c$ because $<$ is transitive. It follows that $a \leq c$.

5. Let $a, b \in A$. If $a < b$ or $a = b$, then $a \leq b$. Otherwise, $b < a$, in which case we have $b \leq a$.

Exercise 3.26:

1. **Yes** because $(\mathbb{N}, >)$ is an ordered set.

2. **No** because $\geq$ is **not** antireflexive (for example, $0 \geq 0$).

3. **Yes** because $(\mathbb{Q}, <)$ is an ordered set.

4. **Yes** because $\{(x, y), (x, z), (y, z)\}$ is antirelfexive, antisymmetric, and transitive.

Exercise 3.27:

1. Let $a \in A$. Since $a = a$, by definition we have $a \leq a$.

2. Let $a, b \in A$ with $a \leq b$ and $b \leq a$. If $a \neq b$, then $a < b$ and $b < a$. But this cannot happen because $<$ is antisymmetric on A ($a < b$ and $b < a$ would imply that $a = b$).

3. Let $a, b, c \in A$ with $a \leq b$ and $b \leq c$. If $a = b$, then we have $a \leq c$ by direct substitution. Similarly, if $b = c$, we have $a \leq c$ by direct substitution. If $a < b$ and $b < c$, then $a < c$ because $<$ is transitive. It follows that $a \leq c$.

Exercise 3.29:

1. **False** because $\{\emptyset, \{\emptyset\}\} = \{0, 1\} = 2$ and $2 \notin 2$ (or equivalently, $\{\emptyset, \{\emptyset\}\} \notin \{\emptyset, \{\emptyset\}\}$).

2. **True** because $\{\emptyset, \{\emptyset\}, \{\emptyset, \{\emptyset\}\}\} = \{0, 1, 2\}$ and $2 \in \{0, 1, 2\}$.

3. **False** because $\{0, 1, 2, 3\} = 4$ and $4 \not<_{\mathbb{N}} 4$.

4. **True** because $(\mathbb{N}, <)$ is an ordered set and $17 \in 18$ is equivalent to $17 < 18$.

Lesson 4

Exercise 4.2:

1. **R is an equivalence relation on A.**

2. **S is not an equivalence relation on A** because $(3, 3) \notin S$, and so, S is not reflexive on A.

3. **T is not an equivalence relation on A** because $(0, 1) \in T$, but $(1, 0) \notin T$, and so, T is not symmetric on A.

4. **U is an equivalence relation on A.**

5. **V is not an equivalence relation on A** because $(0, 1), (1, 2) \in V$, but $(0, 2) \notin V$, and so, V is not transitive on A.

Exercise 4.4:

1. $a - a = 0 = 3 \cdot 0$. So, $3|a - a$. Therefore, $a \equiv_3 a$. So, $\equiv_3$ is reflexive on $\mathbb{Z}$.

2. If $a \equiv_3 b$, then $3|b - a$. So, there is an integer k such that $b - a = 3k$. Then $a - b = 3(-k)$, and so, $3|a - b$. Therefore, $b \equiv_3 a$. So, $\equiv_3$ is symmetric on $\mathbb{Z}$.

3. If $a \equiv_3 b$ and $b \equiv_3 c$, then $3|b - a$ and $3|c - b$. So, there are integers j and k such that $b - a = 3j$ and $c - b = 3k$. Then $c - a = (c - b) + (b - a) = 3k + 3j = 3(k + j)$. So, $3|c - a$. Therefore, $a \equiv_3 c$. So, $\equiv_3$ is transitive on $\mathbb{Z}$.

4. $\equiv_3$ is an equivalence relation on $\mathbb{Z}$ because $\equiv_3$ is reflexive, symmetric, and transitive on $\mathbb{Z}$.

Exercise 4.6:

1. Since $ab = ba$, we see that $(a, b)R(a, b)$, and therefore, R is reflexive on $\mathbb{Z} \times \mathbb{Z}^*$.

2. If $(a, b)R(c, d)$, then $ad = bc$. Therefore, $cb = da$, and so, $(c, d)R(a, b)$. Thus, R is symmetric on $\mathbb{Z} \times \mathbb{Z}^*$.

3. Suppose that $(a, b)R(c, d)$ and $(c, d)R(e, f)$. Then $ad = bc$ and $cf = de$. So, $adcf = bcde$. Therefore, $cd(af - be) = adcf - bcde = 0$. If $a = 0$, then $bc = 0$, and so, $c = 0$ (because $b \neq 0$). So, $de = 0$, and therefore, $e = 0$ (because $d \neq 0$). So, $af = be$ (because they're both 0). If $a \neq 0$, then $c \neq 0$. Therefore, $af - be = 0$, and so, $af = be$. Since $a = 0$ and $a \neq 0$ both lead to $af = be$, we have $(a, b)R(e, f)$. So, R is transitive on $\mathbb{Z} \times \mathbb{Z}^*$.

4. R is an equivalence relation on $\mathbb{Z} \times \mathbb{Z}^*$ because R is reflexive, symmetric, and transitive on $\mathbb{Z} \times \mathbb{Z}^*$.

Exercise 4.8:

1. **True**

2. **True**

3. **False**

4. **False** because $[3] = \{3\}$.

5. **True** because $[1] = \{1, 2\}$ and $[2] = \{1, 2\}$.

6. **False** because $[1] = \{1, 2\}$ and $[3] = \{3\}$.

7. **False** because $[3] = \{3\}$, and so, $|[3]| = 1$.

8. **True** because $[1] \cup [3] = \{1, 2\} \cup \{3\} = \{1, 2, 3\} = A$.

Exercise 4.11:

1. **Not pairwise disjoint**: For example, $\{a, b\} \cap \{b, c\} = \{b\} \neq \emptyset$.

2. **Not pairwise disjoint** because $\{a, b\} \cap \{e, a\} = \{a\} \neq \emptyset$.

3. **Pairwise disjoint**

4. **Not pairwise disjoint**: For example, $6 \in 2\mathbb{Z} \cap 3\mathbb{Z}$.

5. **Not pairwise disjoint**: For example, $4 \in 2\mathbb{Z} \cap (3\mathbb{Z} + 1)$.

6. **Pairwise disjoint**

7. **Pairwise disjoint** because every element of $4\mathbb{Z} + 1$ is odd.

8. **Pairwise disjoint** because every element of $4\mathbb{Z} + 1$ is odd, every element of $4\mathbb{Z} + 3$ is odd, and $(4\mathbb{Z} + 1) \cap (4\mathbb{Z} + 3) = \emptyset$.

Exercise 4.13:

1. **Yes**

2. **No** because $\{a, b\} \cap \{b, c\} = \{b\} \neq \emptyset$, and therefore, X is not pairwise disjoint.

3. **Yes**

4. **Yes**: We can visualize this partition as follows:

$$\mathbb{Z} = \{\ldots, -6, -3, 0, 3, 6, \ldots\} \cup \{\ldots, -5, -2, 1, 4, 7, \ldots\} \cup \{\ldots, -4, -1, 2, 5, 8, \ldots\}$$

5. **Yes**

Exercise 4.17:

1. **False**: $[(36, 36)] = [(0, 0)] = 0$

2. **True**: $[(5, 5)] = [(0, 0)] = 0$

3. **False**: $[(1, 2)] = [(0, 1)] = -1$ and $[(2, 1)] = [(1, 0)] = 1$

4. **True**: $[(3, 4)] = [(0, 1)] = -1$

5. **False**: $[(4, 3)] = [(1, 0)] = 1$

Exercise 4.19:

1. $[(5, 7)] > [(2, 11)]$ because $5 + 11 = 16$, $7 + 2 = 9$, and $16 > 9$.

2. $[(26, 5)] < [(25, 3)]$ because $26 + 3 = 29$, $5 + 25 = 30$, and $29 < 30$.

3. $[(5, 17)] = [(10, 22)]$ because $5 + 22 = 27$ and $17 + 10 = 27$.

4. $[(3, 4)] < [(4, 3)]$ because $3 + 3 = 6$, $4 + 4 = 8$, and $6 < 8$.

Exercise 4.21:

1. **True** because $3 \cdot 1 = 3 \cdot 1$, and so, $\frac{3}{3} = \frac{1}{1} = 1$.

2. **False** because $1 \cdot 1 \neq 7 \cdot 7$.

3. **True** because $(-2)(-14) = 7 \cdot 4$.

Lesson 5

Exercise 5.3:

1. **Not a function from A to B**: requirement 2 is violated because $3 \in A$, but there is no $x \in B$ with $(3, x) \in f$.

2. **This is a function from A to B.**

3. **Not a function from A to B**: requirement 2 is violated because $(2, b), (2, c) \in h$, but $b \neq c$.

4. **This is a function from A to B** (this is a constant function).

5. **Not a function from A to B**: requirement 1 is violated because $m \not\subseteq A \times B$ ($m \subseteq B \times A$).

Exercise 5.5:

1. $f_1 = \{(a, 0), (b, 0)\}$ $\quad$ $f_2 = \{(a, 0), (b, 1)\}$ $\quad$ $f_3 = \{(a, 0), (b, 2)\}$
   $f_4 = \{(a, 1), (b, 0)\}$ $\quad$ $f_5 = \{(a, 1), (b, 1)\}$ $\quad$ $f_6 = \{(a, 1), (b, 2)\}$
   $f_7 = \{(a, 2), (b, 0)\}$ $\quad$ $f_8 = \{(a, 2), (b, 1)\}$ $\quad$ $f_9 = \{(a, 2), (b, 2)\}$

2. $g_1 = \{(0, a), (1, a), (2, a)\}$ $\quad$ $g_2 = \{(0, a), (1, a), (2, b)\}$
   $g_3 = \{(0, a), (1, b), (2, a)\}$ $\quad$ $g_4 = \{(0, a), (1, b), (2, b)\}$
   $g_5 = \{(0, b), (1, a), (2, a)\}$ $\quad$ $g_6 = \{(0, b), (1, a), (2, b)\}$
   $g_7 = \{(0, b), (1, b), (2, a)\}$ $\quad$ $g_8 = \{(0, b), (1, b), (2, b)\}$

Exercise 5.7:

1. $\left(0, \frac{1}{2}, 1, \frac{3}{2}, 2, \frac{5}{2}, 3, \dots\right)$; $\left\{(0,0), \left(1, \frac{1}{2}\right), (2,1), \left(3, \frac{3}{2}\right), (4,2), \left(5, \frac{5}{2}\right), (6,3), \dots\right\}$

2. $f: \mathbb{N} \to \mathbb{N}$, where $f(n) = n + 1$; $\{(0,1), (1,2), (2,3), (3,4), (4,5), (5,6), (6,7), \dots\}$

3. $f: \mathbb{N} \to \mathbb{N}$, where $f(n) = 5n$; $(0, 5, 10, 15, 20, 25, 30, \dots)$

Exercise 5.9:

1. $\left(\frac{n}{2}\right)$; nth term is $\frac{n}{2}$.

2. $(n + 1)$; nth term is $n + 1$.

3. $(5n)$; nth term is $5n$.

Exercise 5.11:

1. $\left(0, \frac{1}{3}, \frac{2}{3}\right)$; $k = 3$

2. $(1, -1, 1, -1, 1, -1, 1, -1, 1, -1, 1)$; $k = 11$

3. $\left(0, \frac{1}{2} + i, 1 + 2i, \frac{3}{2} + 3i, 2 + 4i, \frac{5}{2} + 5i\right)$; $k = 6$

Exercise 5.16:

1. **Injective only**: The range consists of odd integers only. So, for example, $0 \notin \operatorname{ran} f$. Therefore, f is **not** surjective.

2. **Bijective**

3. **Surjective only**: Since $h(1) = 1$ and $h(-1) = 1$, h is **not** injective.

4. **Neither injective nor surjective**: Since $k(1) = \frac{1}{2}$ and $k(-1) = \frac{1}{2}$, k is **not** injective. Since $\frac{1}{n^2+1}$ is always positive, k is **not** surjective (for example, $-1 \notin \operatorname{ran} k$).

Exercise 5.19:

1. $f^{-1} = \{(\square, a), (\Delta, b), (\mho, c)\}$

2. $g^{-1}: 2\mathbb{N} \to \mathbb{N}$, where $g(2n) = n$

3. $h^{-1}: \mathbb{R} \to \mathbb{R}$, where $h^{-1}(r) = \frac{3}{2}(r - 7)$: To see this, let $s = \frac{2}{3}r + 7$ and then switch the roles of r and s to get $r = \frac{2}{3}s + 7$. We now subtract 7 from each side of this equation to get $r - 7 = \frac{2}{3}s$. Now, multiply each side of this last equation by $\frac{3}{2}$ to get $\frac{3}{2}(r - 7) = s$. The result follows.

4. $k^{-1}: \mathbb{C} \to \mathbb{C}$, where $k^{-1}(z) = -iz$: To see this, let $w = iz$ and then switch the roles of z and w to get $z = iw$. We multiply each side of this equation by $-i$ to get $-iw = -i^2 z = -(-1)z = z$. The result follows.

Exercise 5.21:

1. Let $f, g \in {}^A2$ with $f \neq g$. Since f and g are different, there is some $a \in A$ such that either $f(a) = 0, g(a) = 1$ or $f(a) = 1, g(a) = 0$.

   First assume that $f(a) = 0$ and $g(a) = 1$. Since $f(a) = 0$, $a \notin F(f)$. Since $g(a) = 1$, $a \in F(g)$. So, $F(f) \neq F(g)$.

   Now assume that $f(a) = 1$ and $g(a) = 0$. Since $f(a) = 1$, $a \in F(f)$. Since $g(a) = 0$, $a \notin F(g)$. So, $F(f) \neq F(g)$.

   Since $f \neq g$ implies $F(f) \neq F(g)$, F is injective.

2. To see that F is surjective, let $C \in \mathcal{P}(A)$, so that $C \subseteq A$. Define $f \in {}^A2$ by $f(x) = \begin{cases} 0 & \text{if } x \notin C. \\ 1 & \text{if } x \in C. \end{cases}$ Then $x \in F(f)$ if and only if $f(x) = 1$ if and only if $x \in C$. So, $F(f) = C$. Since $C \in \mathcal{P}(A)$ was arbitrary, F is surjective.

3. Since F is injective (by 1 above) and F is surjective (by 2 above), it follows that F is bijective.

4. As in Example 5.20, the inverse of F is the function $F^{-1}: \mathcal{P}(A) \to {}^A B$ defined by $F^{-1}(C)(x) = \begin{cases} 0 & \text{if } x \notin C. \\ 1 & \text{if } x \in C. \end{cases}$

Exercise 5.23:

1. If n is even, then there is an integer k such that $n = 2k$. So, $f(n) = f(2k) = \frac{2k}{2} = k$. Therefore, $(g \circ f)(n) = g(f(n)) = g(k) = \mathbf{0}$.

2. If n is odd, then there is an integer k such that $n = 2k + 1$. So, $f(n) = f(2k + 1) = \frac{2k+1}{2}$. Therefore, $(g \circ f)(n) = g(f(n)) = g\left(\frac{2k+1}{2}\right) = \mathbf{1}$.

3. $(g \circ f)(n) = \begin{cases} \mathbf{0} & \text{if } \mathbf{n} \text{ is even.} \\ \mathbf{1} & \text{if } \mathbf{n} \text{ is odd.} \end{cases}$

Exercise 5.25:

1. **Cauchy sequence**; converging to 1; **rational number**

2. **Not a Cauchy sequence**

3. **Not a Cauchy sequence**

4. **Cauchy sequence**; converging to 0; **rational number**

5. **Cauchy sequence**; converging to π; **not a rational number**

Lesson 6

Exercise 6.2:

1. $A \sim B$ via the bijection $\{(a,x),(b,y),(c,z),(d,w)\}$.

2. $A \nsim B$ because $|A| = 101$ and $|B| = 102$.

3. $A \nsim B$ because A is finite and B is infinite.

4. $A \sim B$ via the bijection $f : \mathbb{Z} \to 2\mathbb{Z}$ defined by $f(n) = 2n$.

5. $A \sim B$ via the bijection $F : {}^{\mathbb{N}}2 \to \mathcal{P}(\mathbb{N})$ defined by $F(f) = \{n \in \mathbb{N} \mid f(n) = 1\}$ (see Exercise 5.21).

6. $A \sim B$ via the bijection $f : [0,1] \to [0,3]$ defined by $f(x) = 3x$.

Exercise 6.3:

1. For any set A, the identity function $i_A : A \to A$ is a bijection, showing that $\sim$ is reflexive.

2. For sets A and B, if $f : A \cong B$, then $f^{-1} : B \cong A$, showing that $\sim$ is symmetric.

3. For sets A, B, and C, if $f : A \cong B$ and $g : B \cong C$, then $g \circ f : A \cong C$ by Composition Fact 3 (see Lesson 5), showing that $\sim$ is transitive.

Exercise 6.5:

1. $1, 3, 5, 7, \ldots$

2. $-1, 1, -3, 3, -5, 5, \ldots$

3. $0, -1, 1, -2, 2, -\frac{1}{2}, \frac{1}{2}, -3, 3, -\frac{1}{3}, \frac{1}{3}, -4, 4, -\frac{3}{2}, \frac{3}{2}, -\frac{2}{3}, \frac{2}{3}, -\frac{1}{4}, \frac{1}{4}, -5, 5, -\frac{1}{5}, \frac{1}{5}, \ldots$

Note: For part 3 of Exercise 6.5, use part 4 of Example 6.1 and part 4 of Example 6.4 as a guide.

Exercise 6.7: There are infinitely many solutions to this exercise. I will provide one solution here.

1. $B = \{\{a\} \mid a \in A\}$
2. $f(a) = \{a\}$

Exercise 6.9:

1. $A \prec B$ because $|A| = 2$ and $|B| = 2^2 = 4$.

2. $A \sim B$ because $|A| = 2$ and $|B| = 2^1 = 2$.

3. $A \prec B$ because A is finite and B is infinite.

4. $A \prec B$ by Equinumerosity Fact 1 (Cantor's Theorem).

5. $A \prec B$ because $\mathbb{Z} \prec \mathcal{P}(\mathbb{Z})$ (by Equinumerosity Fact 1) and $\mathbb{Z} \sim \mathbb{N}$, so that $\mathcal{P}(\mathbb{Z}) \sim \mathcal{P}(\mathbb{N})$.

6. $A \prec B$ because $\mathcal{P}(\mathbb{Q}) \prec \mathcal{P}(\mathcal{P}(\mathbb{Q}))$ (by Equinumerosity Fact 1) and $\mathcal{P}(\mathcal{P}(\mathbb{Q})) \sim \mathcal{P}(\mathcal{P}(2\mathbb{Z}))$ (because $\mathbb{Q} \sim 2\mathbb{Z}$).

Exercise 6.11:

1. Since $A \subseteq B$, it follows that $A \preccurlyeq B$ (define $f: A \to B$ by $f(a) = a$ for all $a \in A$).

2. Define $g: B \to A$ by $g(x) = \frac{1}{3}x + 2$. Then $g(0) = 2$ and $g(6) = 4$. So, $g: [0,6] \to [2,4]$ is a bijection. Since $[2,4] \subseteq (1,5)$, we have $[0,6] \preccurlyeq (1,5)$.

3. Since $A \preccurlyeq B$ and $B \preccurlyeq A$, by the Cantor-Schroeder-Bernstein Theorem, $A \sim B$.

Lesson 7

Exercise 7.3:

1. This is a statement. This statement happens to be false.

2. This is **not** a statement. It is a **question**.

3. This is a statement. This statement happens to be true.

4. This is **not** a statement. It is a **command**.

5. This is a statement. Either Andrea is not feeling well today (in which case the statement is true) or she is feeling well today (in which case the statement is false).

Exercise 7.6:

1. This is an **atomic statement**.

2. This is a **compound statement**. It uses the logical connective "or."

3. This is a **compound statement**. It uses the logical connective "not."

4. This is an **atomic statement**. Even though the word "and" appears in the statement, here it is part of the name of the book. It is not being used as a logical connective.

5. This is a **compound statement**. It uses the logical connective "if...then."

6. This is an **atomic statement**.

7. This is a **compound** statement. It actually uses two connectives: "or" and "not."

8. This is a **compound statement**. It uses the logical connective "if and only if."

9. This is an **atomic statement**. Even though the word "not" appears in the statement, it is **not** being used as a logical connective.

10. This is a **compound statement**. Like part 7 above, it uses two connectives: "and" and "not." Note that in sentential logic the word "but" has the same meaning as the word "and." In English, the word "but" is used to introduce contrast with the part of the sentence that has already been mentioned. However, logically it is no different from "and."

Exercise 7.10: There are **eight** possible truth assignments for this list of propositional variables. We can visualize this list of truth assignments with the following table:

p	q	r
T	T	T
T	T	F
T	F	T
T	F	F
F	T	T
F	T	F
F	F	T
F	F	F

Exercise 7.12:

1. $p \wedge q \equiv T \wedge T \equiv \mathbf{T}$.
2. $p \wedge q \equiv F \wedge F \equiv \mathbf{F}$.
3. $p \wedge q \equiv F \wedge T \equiv \mathbf{F}$.

p	q	$p \wedge q$
T	T	T
T	F	F
F	T	F
F	F	F

Exercise 7.13:

1. $p \vee q \equiv T \vee T \equiv \mathbf{T}$.
2. $p \vee q \equiv F \vee F \equiv \mathbf{F}$.
3. $p \vee q \equiv T \vee F \equiv \mathbf{T}$.
4. $p \vee q \equiv F \vee T \equiv \mathbf{T}$.

p	q	$p \vee q$
T	T	T
T	F	T
F	T	T
F	F	F

Exercise 7.14:

1. $p \rightarrow q \equiv T \rightarrow T \equiv \mathbf{T}$.
2. $p \rightarrow q \equiv F \rightarrow F \equiv \mathbf{T}$.
3. $p \rightarrow q \equiv T \rightarrow F \equiv \mathbf{F}$.
4. $p \rightarrow q \equiv F \rightarrow T \equiv \mathbf{T}$.

p	q	$p \rightarrow q$
T	T	T
T	F	F
F	T	T
F	F	T

Exercise 7.15:

1. $p \leftrightarrow q \equiv T \leftrightarrow T \equiv \mathbf{T}$.
2. $p \leftrightarrow q \equiv F \leftrightarrow F \equiv \mathbf{T}$.
3. $p \leftrightarrow q \equiv T \leftrightarrow F \equiv \mathbf{F}$.
4. $p \leftrightarrow q \equiv F \leftrightarrow T \equiv \mathbf{F}$.

p	q	$p \leftrightarrow q$
T	T	T
T	F	F
F	T	F
F	F	T

Exercise 7.17: $\neg p \equiv \neg F \equiv T.$ ←——————

p	$\neg p$
T	F
F	T

Exercise 7.19: Note that p and q are both false.

1. $p \to q$ represents **"If frogs are birds, then 2 < 1."** Since p is false, $p \to q$ is **true**.

2. $\neg p \lor q$ represents the statement **"Frogs are not birds or 2 < 1."** Since $\neg p$ is true, $\neg p \lor q$ is **true**. Note once again that $\neg p \lor q$ always means $(\neg p) \lor q$. In general, without parentheses present, we always apply negation before any of the other connectives.

3. $p \leftrightarrow q$ represents **"Frogs are birds if and only if 2 < 1."** Since p and q are both false, $p \leftrightarrow q$ is **true**.

4. $(p \to q) \land (q \to p)$ represents **"If frogs are birds, then 2 < 1 and if 2 < 1, then frogs are birds."** Since p is false, $p \to q$ is true. Since q is false, $q \to p$ is true. Since $p \to q$ and $q \to p$ are both true, $(p \to q) \land (q \to p)$ is **true**.

5. $\neg(p \land q)$ represents the statement **"It is not the case that both frogs are birds and 2 < 1."** Since p and q are both false, $p \land q$ is false. It follows that $\neg(p \land q)$ is **true**.

6. $\neg p \lor \neg q$ represents the statement **"Frogs are not birds or 2 is not less than 1."** Since p and q are both false, $\neg p$ and $\neg q$ are both true. It follows that $\neg p \lor \neg q$ is **true**.

Exercise 7.21:

1.

p	q	r	$\neg r$	$q \land \neg r$	$p \leftrightarrow (q \land \neg r)$
T	T	T	F	F	F
T	T	F	T	T	T
T	F	T	F	F	F
T	F	F	T	F	F
F	T	T	F	F	T
F	T	F	T	T	F
F	F	T	F	F	T
F	F	F	T	F	T

2. We use the highlighted row in the truth table above to get a truth value of **false**.

3. $p \leftrightarrow (q \land \neg r) \equiv T \leftrightarrow (q \land \neg T) \equiv T \leftrightarrow (q \land F) \equiv T \leftrightarrow F \equiv F.$ So, the truth value **can be determined** and it is **false**.

Exercise 7.23: We will show that the truth tables for $p \to q$ and $\neg p \lor q$ are the same.

p	q	$p \to q$	$\neg p$	$\neg p \lor q$
T	T	T	F	T
T	F	F	F	F
F	T	T	T	T
F	F	T	T	T

Exercise 7.25: We will show that the truth tables for $p \to q$ and $\neg q \to \neg p$ are the same.

p	q	$p \to q$	$\neg p$	$\neg q$	$\neg q \to \neg p$
T	T	T	F	F	T
T	F	F	F	T	F
F	T	T	T	F	T
F	F	T	T	T	T

Exercise 7.27: We will show that the truth tables for $\neg(p \lor q)$ and $\neg p \land \neg q$ are the same.

p	q	$\neg p$	$\neg q$	$p \lor q$	$\neg(p \lor q)$	$\neg p \land \neg q$
T	T	F	F	T	F	F
T	F	F	T	T	F	F
F	T	T	F	T	F	F
F	F	T	T	F	T	T

Exercise 7.30:

$$[(\neg p \lor q) \land p] \lor q \equiv [p \land (\neg p \lor q)] \lor q \equiv [(p \land \neg p) \lor (p \land q)] \lor q \equiv [F \lor (p \land q)] \lor q$$
$$\equiv [(p \land q) \lor F] \lor q \equiv (p \land q) \lor q \equiv (q \land p) \lor q \equiv q$$

So, we see that $[(\neg p \lor q) \land p] \lor q$ is logically equivalent to the atomic statement q.

Notes: (1) For the first equivalence, we used the first commutative law.

(2) For the second equivalence, we used the first distributive law.

(3) For the third equivalence, we used the first negation law.

(4) For the fourth equivalence, we used the second commutative law.

(5) For the fifth equivalence, we used the fourth identity law.

(6) For the sixth equivalence, we used the first commutative law.

(7) For the last equivalence, we used the second absorption law.

Exercise 7.32: We will show that the final column of the truth table for $(p \to q) \leftrightarrow (\neg q \to \neg p)$ consists of only the truth value T.

p	q	$\neg p$	$\neg q$	$p \to q$	$\neg q \to \neg p$	$(p \to q) \leftrightarrow (\neg p \to \neg q)$
T	T	F	F	T	T	T
T	F	F	T	F	F	T
F	T	T	F	T	T	T
F	F	T	T	T	T	T

Exercise 7.34: If $p \equiv T$, then $p \wedge \neg p \equiv T \wedge F \equiv F$. If $p \equiv F$, then $p \wedge \neg p \equiv F \wedge T \equiv F$. Since both possible truth assignments of the propositional variable p lead to the statement $p \wedge \neg p$ having truth value F, it follows that $p \wedge \neg p$ is a contradiction.

Exercise 7.36:

1. **True**

2. **True** (let $x = 0$ and $y = 2$)

3. **False** (for example, let $x = 0$ and $y = 1$)

4. **True**

5. **False** (given y, let $x = y$)

Exercise 7.38:

1. $\neg \forall x (x = x) \equiv \exists x \neg (x = x) \equiv \exists x (x \neq x)$

2. $\neg \exists x (x < x) \equiv \forall x \neg (x < x) \equiv \forall x (x \geq x)$

3. $\neg \forall x \forall y (x = y \wedge x \in y) \equiv \exists x \exists y \neg (x = y \wedge x \in y) \equiv \exists x \exists y (\neg (x = y) \vee \neg (x \in y))$

   $\equiv \exists x \exists y (x = y \to \neg (x \in y)) \equiv \exists x \exists y (x = y \to x \notin y)$

4. $\neg \forall x \exists y (x \leq x \to y = x) \equiv \exists x \forall y \neg (x \leq x \to y = x) \equiv \exists x \forall y \neg (\neg (x \leq x) \vee y = x)$

   $\equiv \exists x \forall y (x \leq x \wedge \neg (y = x)) \equiv \exists x \forall y (x \leq x \wedge y \neq x)$

5. $\neg \exists x \forall y \exists z ((x \in y \vee y \in z) \to x \in z) \equiv \forall x \exists y \forall z \neg ((x \in y \vee y \in z) \to x \in z)$

   $\equiv \forall x \exists y \forall z \neg (\neg (x \in y \vee y \in z) \vee x \in z) \equiv \forall x \exists y \forall z ((x \in y \vee y \in z) \wedge \neg (x \in z))$

   $\equiv \forall x \exists y \forall z ((x \in y \vee y \in z) \wedge x \notin z)$

Exercise 7.40:

1. **True**

2. **True**

3. **False**

4. **True**

Exercise 7.42: We start with the atomic formulas $z \in x$, $y \in x$, $w \in y$, and $w \in x$.

We apply the unary connective $\neg$ to the atomic formula $w \in x$ to get the formula $\neg w \in x$, or equivalently, $w \notin x$.

We apply the binary connective $\wedge$ to the atomic formula $w \in y$ and the formula $w \notin x$ to get the formula $w \in y \wedge w \notin x$.

We apply the quantifier $\forall w$ to this last formula to get the formula $\forall w(w \in y \wedge w \notin x)$.

We apply the unary connective $\neg$ to this last formula to get the formula $\neg \forall w(w \in y \wedge w \notin x)$.

We apply the binary connective $\wedge$ to the atomic formula $y \in x$ and the last formula to get the formula $y \in x \wedge \neg \forall w(w \in y \wedge w \notin x)$.

We apply the quantifier $\forall y$ to this last formula to get the formula $\forall y(y \in x \wedge \neg \forall w(w \in y \wedge w \notin x))$.

We apply the unary connective $\neg$ to this last formula to get the formula

$$\neg \forall y(y \in x \wedge \neg \forall w(w \in y \wedge w \notin x)).$$

We apply the quantifier $\exists z$ to the atomic formula $z \in x$ to get the formula $\exists z(z \in x)$.

We apply the binary connective $\wedge$ to the formulas $\exists z(z \in x)$ and $\neg \forall y(y \in x \wedge \neg \forall w(w \in y \wedge w \notin x))$ to get the formula $\exists z(z \in x) \wedge \neg \forall y(y \in x \wedge \neg \forall w(w \in y \wedge w \notin x))$.

Finally, we apply the quantifier $\forall x$ to this last formula to get the formula

$$\forall x \left(\exists z(z \in x) \wedge \neg \forall y(y \in x \wedge \neg \forall w(w \in y \wedge w \notin x)) \right).$$

Exercise 7.44:

1. **Not a sentence**: x and y are both free.

2. **Sentence**

3. **Not a sentence**: The first instance of y is free and the second instance of x is free.

4. **Not a sentence**: The second instance of x is free.

5. **Sentence**

6. **Not a sentence**: The third instance of x is free.

Exercise 7.46: Given sets x and y, by the pairing axiom, we get the set $\{x, y\}$. Again, using the pairing axiom, we get the set $\{x, x\} = \{x\}$. Using the pairing axiom one more time, we get the set $\{\{x\}, \{x, y\}\}$, which is equal to the ordered pair (x, y) (this definition can be found at the beginning of Lesson 3).

Exercise 7.48:

1. If we let $x = \{4, \{4\}\}$ (this is a set by part 2 of Example 7.47 and the Pairing Axiom), the union axiom gives us the set $\cup x = \cup\{4, \{4\}\} = 4 \cup \{4\} = \{0, 1, 2, 3\} \cup \{4\} = \{0, 1, 2, 3, 4\} = 5$.

2. By the Pairing Axiom, $\{n, n\} = \{n\}$ is a set. Then by the Union Axiom, we get the set

$$\cup\{n, \{n\}\} = n \cup \{n\} = \{0, 1, \ldots, n - 1\} \cup \{n\} = \{0, 1, \ldots, n - 1, n\} = n + 1.$$

Exercise 7.50: $\{0, 1, 2, 3, \{1\}, \{2\}, \{0, 2\}, \{1, 2\}\} = \mathcal{P}(3)$, which is a set by the Power Set Axiom.

Exercise 7.51: First note that $\emptyset \in x$ is an abbreviation for $\exists y\big(y \in x \land \forall z(z \notin y)\big)$.

Also, $y \cup \{y\} \in x$ is an abbreviation for $\exists z\Big(z \in x \land \forall w\big(w \in z \leftrightarrow (w \in y \lor w = y)\big)\Big)$.

So, the fully unabbreviated form of this axiom is the following:

$$\exists x \bigg(\exists y\big(y \in x \land \forall z(z \notin y)\big) \land \forall y \Big(y \in x \to \exists z \big(z \in x \land \forall w(w \in z \leftrightarrow (w \in y \lor w = y))\big)\Big)\bigg)$$

Exercise 7.53: We can define $\cap b = \{w \in \cup b \mid \forall z(z \in b \land w \in z)\}$. Here, we have used the formula $\phi(w, y, x)$ defined by $\forall z(z \in y \land w \in z)$, together with the parameter b. The bounding set is $\cup b$, which we know is a set by the Union Axiom.

Lesson 8

Exercise 8.2:

1. **Yes**

2. **Yes**

3. **Yes**

4. **Yes**

5. $0 <^* 1 <^* 2 <^* 3 <^* 4 <^* 5 <^* \cdots <^* \star$

Exercise 8.4:

1. $(4, \in) \cong (\{a, b, c, d\}, <)$ via the isomorphism $f = \{(0, a), (1, b), (2, c), (3, d)\}$. We can visualize this isomorphism as follows:

$$0 < 1 < 2 < 3$$
$$\downarrow \quad \downarrow \quad \downarrow \quad \downarrow$$
$$a < b < c < d$$

2. $(5, \in) \ncong (6, \in)$. Since 5 and 6 have different cardinalities, there can be no bijection between them.

3. $(\mathbb{N}, <') \cong (A, <^*)$ via the isomorphism $f(n) = \begin{cases} n - 1 & \text{if } n \neq 0. \\ \star & \text{if } n = 0. \end{cases}$ We can visualize this isomorphism as follows:

$$1 <' 2 <' 3 <' 4 <' 5 <' \cdots <' 0$$
$$\downarrow \quad \downarrow \quad \downarrow \quad \downarrow \quad \downarrow \qquad\qquad \downarrow$$
$$0 <^* 1 <^* 2 <^* 3 <^* 4 <^* \cdots <^* \star$$

4. $(500, \in) \ncong (\omega, \in)$. Since 5 is finite and ω is infinite, there can be no bijection between them.

5. $(\omega, \in) \ncong (\mathbb{N}, <')$. If $f: \omega \to \mathbb{N}$ is any bijection, then there is $n \in \omega$ with $f(n) = 0$. Since $n < n + 1$, we must have $f(n) <' f(n + 1)$, or equivalently, $0 <' f(n + 1)$. This is impossible because there is no element greater than 0 in the ordering $<'$.

Exercise 8.6:

1. $[0, 8) = \{0, 2, 4, 6\}$

2. $[0, 5) = \{0, 2, 4, 6, 8, \dots, 1, 3\} = 2\mathbb{N} \cup \{1, 3\}$

3. $\mathrm{pred}(\mathbb{N}, 0) = \varnothing$

4. $\mathrm{pred}(\mathbb{N}, 14) = \{0, 2, 4, 6, 8, 10, 12\}$

5. $\mathrm{pred}(\mathbb{N}, 15) = \{0, 2, 4, 6, 8, \dots, 1, 3, 5, 7, 9, 11, 13\} = \mathbb{N} \cup \{1, 3, 5, 7, 9, 11, 13\}$

6. $f: \omega \to \mathbb{N}$ can be defined by $f(n) = 2n$

7. $f[\omega] = 2\mathbb{N}$

Exercise 8.8:

1. $\{a\}$ is **not** necessarily transitive. For example, $\{\varnothing\}$ is transitive, but $\{\{\varnothing\}\}$ is not $(\varnothing \in \{\varnothing\} \in \{\{\varnothing\}\}$, but $\varnothing \notin \{\{\varnothing\}\}$.

2. $\mathcal{P}(a)$ **is transitive.** To see this, let $b \in \mathcal{P}(a)$ and let $c \in b$. By the definition of $\mathcal{P}(a)$, $b \subseteq a$. Since $c \in b$ and $b \subseteq a$, we have $c \in a$. Since a is transitive, $c \subseteq a$. Thus, $c \in \mathcal{P}(a)$. Since $c \in b$ was arbitrary, $b \subseteq \mathcal{P}(a)$.

3. (a, a) is **not** necessarily transitive. For example, $\varnothing$ is transitive, but $(\varnothing, \varnothing) = \{\{\varnothing\}\}$ is not. The explanation is the same as in part 1 above.

4. $\bigcup a$ **is transitive.** To see this, let $b \in \bigcup a$. Then $b \in x$ for some $x \in a$. Since a is transitive, $b \in a$. By Problem 62 from Problem Set 2, $b \subseteq \bigcup a$.

Exercise 8.10:

1. **4**

2. $\omega + 1$

3. $\omega + \omega$

Exercise 8.12: No

$\omega + 1 = \{0, 1, 2, \dots, \omega\}$ is the successor of ω. This ordering looks like the natural numbers followed by one additional element that is larger than all the natural numbers. We can visualize this by placing ω almonds followed by one berry.

$$\diamond \diamond \diamond \diamond \diamond \diamond \diamond \diamond \diamond \diamond \diamond \diamond \ \cdots \ \circ$$

On the other hand, $1 + \omega = \bigcup\{1 + n \mid n \in \omega\} = \omega$, as shown in part 2 of Example 8.11.

Exercise 8.14:

$5\omega + 3 = \omega + 3$

$\omega \cdot 5 + 3 = \omega + \omega + \omega + \omega + \omega + 3$

$3 + 5\omega = 3 + \omega = \omega$

$3 + \omega \cdot 5 = 3 + (\omega + \omega + \omega + \omega + \omega) = \omega + \omega + \omega + \omega + \omega$

So, we have

$$3 + 5\omega \in 5\omega + 3 \in 3 + \omega \cdot 5 \in \omega \cdot 5 + 3.$$

Exercise 8.17:

1. **5 is a cardinal** because all natural numbers are cardinals (see part 1 of Example 8.16).

2. **$\omega + 1$ is not a cardinal** because it is equinumerous with the smaller ordinal ω. We can define a bijection $f\colon \omega + 1 \to \omega$ by $f(\alpha) = \begin{cases} \alpha + 1 & \text{if } \alpha \in \omega. \\ 0 & \text{if } \alpha = \omega. \end{cases}$

3. **$1 + \omega$ is a cardinal** because it is equal to ω, which is a cardinal (see part 2 of Example 8.16).

4. **$\omega \cdot 2$ is not a cardinal** because it is equinumerous with the smaller ordinal ω. We can define a bijection $f\colon \omega \cdot 2 \to \omega$ by $f(\alpha) = \begin{cases} 2\alpha & \text{if } \alpha \in \omega. \\ 2k + 1 & \text{if } \alpha = \omega + k. \end{cases}$

5. **2ω is a cardinal** because it is equal to ω, which is a cardinal (see part 2 of Example 8.16).

Exercise 8.20:

1. Assume toward contradiction that $f\colon \omega_\omega \to \alpha$ is a bijection for some $\alpha \in \omega_\omega$. Then there is $n \in \omega$ with $\alpha \in \omega_n$. So, $|\omega_\omega| \leq \alpha < \omega_n$, and therefore, $|\omega_\omega| < \omega_n$. Since $\omega_n \subseteq \omega_\omega$, $|\omega_n| \leq |\omega_\omega|$. So, $|\omega_n| < \omega_n$, contradicting that ω_n is a cardinal.

2. Assume toward contradiction that ω_ω is a successor cardinal. Then there is $n \in \omega$ with $\omega_\omega = \omega_n^+$. Then we have $|\omega_\omega| = |\omega_n^+| = \omega_n^+ = \omega_{n+1} < \omega_{n+2}$. So, $|\omega_\omega| < \omega_{n+2}$. Since $\omega_{n+2} \subseteq \omega_\omega$, $|\omega_{n+2}| \leq |\omega_\omega|$. So, $|\omega_{n+2}| < \omega_{n+2}$, contradicting that ω_{n+2} is a cardinal. So, ω_ω is a limit cardinal.

Exercise 8.22: Define $f\colon \kappa \to (\kappa \times \{0\}) \cup (\kappa \times \{1\})$ by $f(\alpha) = (\alpha, 0)$. Then f is injective. It follows that $\kappa \leq |(\kappa \times \{0\}) \cup (\kappa \times \{1\})| = \kappa + \kappa$.

Since $(\kappa \times \{0\}) \cup (\kappa \times \{1\}) \subseteq \kappa \times \kappa$, we have

$$\kappa + \kappa = |(\kappa \times \{0\}) \cup (\kappa \times \{1\})| \leq |\kappa \times \kappa| = \kappa \cdot \kappa = \kappa \text{ (by Cardinal Fact 1)}.$$

Since $\kappa \leq \kappa + \kappa$ and $\kappa + \kappa \leq \kappa$, by the Cantor-Schroeder-Bernstein Theorem, $\kappa + \kappa = \kappa$.

Exercise 8.24: Since $(2^\omega)^\omega = 2^{\omega \cdot \omega} = 2^\omega$, by letting $\kappa = 2^\omega$ in König's Theorem, we see that $cf(2^\omega)$ cannot be equal to ω. In particular, $2^\omega \neq \omega_\omega$.

INDEX

About the Author

Dr. Steve Warner, a New York native, earned his Ph.D. at Rutgers University in Pure Mathematics in May 2001. While a graduate student, Dr. Warner won the TA Teaching Excellence Award.

After Rutgers, Dr. Warner joined the Penn State Mathematics Department as an Assistant Professor and in September 2002, he returned to New York to accept an Assistant Professor position at Hofstra University. By September 2007, Dr. Warner had received tenure and was promoted to Associate Professor. He has taught undergraduate and graduate courses in Precalculus, Calculus, Linear Algebra, Differential Equations, Mathematical Logic, Set Theory, and Abstract Algebra.

From 2003 – 2008, Dr. Warner participated in a five-year NSF grant, "The MSTP Project," to study and improve mathematics and science curriculum in poorly performing junior high schools. He also published several articles in scholarly journals, specifically on Mathematical Logic.

Dr. Warner has nearly two decades of experience in general math tutoring and tutoring for standardized tests such as the SAT, ACT, GRE, GMAT, and AP Calculus exams. He has tutored students both individually and in group settings.

In February 2010 Dr. Warner released his first SAT prep book "The 32 Most Effective SAT Math Strategies," and in 2012 founded Get 800 Test Prep. Since then Dr. Warner has written books for the SAT, ACT, SAT Math Subject Tests, AP Calculus exams, and GRE. In 2018 Dr. Warner released his first pure math book called "Pure Mathematics for Beginners." Since then he has released several more books, each one addressing a specific subject in pure mathematics.

Dr. Steve Warner can be reached at

steve@SATPrepGet800.com

BOOKS BY DR. STEVE WARNER